THE UNKNOWN ADVENTURER

THE UNKNOWN ADVENTURER

A STORY OF DREAMING BIG, LIVING BIG, AND LEARNING FROM FAILURE

BARRY WALTON

© 2024 by Barry Walton
THE UNKNOWN ADVENTURER
A Story of Dreaming Big, Living Big, and Learning From Failure

All rights reserved solely by the author. The author guarantees all contents are original and do not infringe upon the legal rights of any other person or work. No part of this book may be reproduced in any form without the permission of the author.

Printed in the United States of America.

ISBN-13: 979-8-218-53663-3
HC ISBN-13: 979-8-218-53684-8
Library of Congress Control Number: 2024922165

Printed in the United States of America.
First printing edition 2024

Endless Publishing
Rockledge, Florida

"Through the lens of his camera, Barry would weave a cinematic tapestry—two documentaries in all—that would immortalize the madness of The High."

—Rajat Chauhan,
Race Director of La Ultra The High

"Barry Walton is a true adventurer. In the Himalayas, I witnessed him relentlessly pursuing his craft, trying to get "the shot," whether that was in a crushing snow storm at 18,000 feet or on the back of a motorcycle . . . his passion for capturing and recording humans testing their potential in extreme circumstances is remarkable."

—Molly Sheridan, Author & Ultra Runner

"Looking back, I realize that during the filming of *The Violin Ghetto*, Barry was trying to uncover the meaning behind the most radical and courageous life choices . . . and to pursue their passions."

—Giorgio Priori, Director of *The Violin Ghetto*

"Most of my youth was living the next house down from Barry Walton . . . the incredible imagination and extremely colorful creativity I saw constantly radiating out of the enormous personality of Barry was inspiring and has impacted me in significant ways throughout my entire life so many decades later."

—Brian Leman, Childhood Friend, Marine

"The Three Amigos, Mom, Barry, and I. Always laughing and smiling . . . getting side-eyed by the other family members . . . a big cousin moment."

—Kim Short, Cousin & Mother

TABLE OF CONTENTS

PROLOGUE XIII
CHAPTER 1 1
CHAPTER 2 15
CHAPTER 3 31
CHAPTER 4 47
CHAPTER 5 67
CHAPTER 6 87
CHAPTER 7 115
CHAPTER 8 129
CHAPTER 9 147
CHAPTER 10 155
CHAPTER 11 175
CHAPTER 12 205
CHAPTER 13 227

DEDICATION

To my wife, my son David, my father, mother, and everyone who believed in my audacious, outrageous, and oversized dreams. Dream big!

Inspired by my fifth-grade teacher, Mrs. Winters, who pushed me to write my first book at nine years old called, The First Boy in Space. And to all teachers, we need you.

"THE ROAD NOT TAKEN"

By Robert Frost

Two roads diverged in a yellow wood,
And sorry I could not travel to both
And be one traveler, long I stood
And looked down one as far as I could
To where it bent in the undergrowth;

Then took the other, as just as fair,
And having perhaps the better claim,
Because it was grassy and wanted wear;
Though as for that the passing there
Had worn them really about the same,

And both that morning equally lay
In leaves no step had trodden black.
Oh, I kept the first for another day!
Yet knowing how way leads on to way,
I doubted if I should ever come back.

I shall be telling this with a sigh
Somewhere ages and ages hence:
Two roads diverged in a wood, and I—
I took the one less traveled by,
And that has made all the difference.

PROLOGUE

I have no business writing this book. I don't even belong in the ranks of those bearing the title of author. Based on my experience in academia, a respectable list of English teachers considered me to be a failure; therefore, even the institutions of education would not have me as a member. But I do have a story I want to share with you about how to dare venture into the unknown, have your dreams fulfilled, and return with something to share with the world.

In my nearly fifty years on this planet, I have traveled by car across the USA from coast to coast; Mexico from border to border, flown by plane to faraway places, and traveled at sea to worlds I never knew existed. Amidst these life passages, I have worked with some of the biggest movie stars of our time. I have won an Emmy, made commercials, and won awards in documentary films that I never thought possible.

In my pursuit of finding the path to a greater sense of reality, I have chased dreams that have skirted dangerously close to something akin to a blaze of fire, and I have sat on other dreams that never left the couch in my living room. I have been let down, disappointed, used, abused, misguided, misinformed and disenfranchised more times than I can count. However, time and time again, I got back up and plodded on in search of my dreams. I persistently held onto a flickering beacon of hope even when finding myself alone, scared and running away from the things I wanted most. However, with time, I overcame my fears and faced my dragons. Slowly, I began to feel the success that comes from hard work, vulnerability, and risk. Through it all, I have learned some of the most important lessons of my life and uncovered the skills needed to succeed, which I hope to share with you here.

It wasn't easy. However, through perseverance, I survived the depths of an insecure start in venturing into the unknown, and no matter how hard it got, I pushed on. I believed that I could dream big and these dreams, no matter how immense, could come true. I held onto that belief like clinging to driftwood from a shipwreck. Through struggle and perseverance, I began to grasp the practical skills that would help me survive and ultimately thrive in a very competitive and sometimes cruel world. These skills have given me the power to demystify my perception of the past and future in my life and have given birth to a vision of my life today.

It has been the process of understanding how to transform my dreams into reality that has led to great adventures—to places I never thought I would see. And it has helped me to reach a deeper understanding of life. Each time I ventured forth, I returned with one more piece of the puzzle that I didn't have before. Each of these pieces has given me insight into how to live life and wisdom in what things I should leave behind.

I am now arguably closer to the end of my life than I am to the start. I am more or less as equally known and unknown by the world as I was at my birth. I don't feel special, nor do I feel gifted. I am very normal and feel unqualified for the task ahead. Taking this into consideration, I have no reason to think writing a book will be a lucrative investment nor a great use of the limited time I have left on this planet. Rest assured that I am not trying to fix people nor bore them with a list of cliché sayings and statements that will make you feel better about yourself. The lessons in this book are not magical; they're not well structured but merely the methods by which I have learned to live life—methods that, if applied appropriately, may help you find a pathway into making your dreams materialize.

Now, after years of collecting all my experiences into one place, I have worked to condense in writing what I have learned into an entertaining and informative message, which I hope is a path that anyone can follow.

It is my wish that reading this book will illuminate your perspective on a life that only you can envision by venturing into places and spaces in your existence yet undiscovered. In the process, you can hopefully overcome your fears and return with a new perspective on life—one that would not have otherwise been revealed. The road ahead may not be easy, but if you choose to go forward and face the unknown, you will return with new skills and an understanding of life that you have not had prior. So, I encourage you to go forth, overcome your fears, achieve your dreams, and return with something that you can give back to the world.

Mom, Kevin (brother), myself and Marissa (sister) standing in front of the Trenton Hills School bus that Mom drove us to school in (1982)

*"Time heals all wounds,
but it doesn't hide the scars."*

CHAPTER 1

THE BIRTH OF ADVENTURE

TAYLOR FALLS, MINNESOTA
THE ADVENTURE JUMPS OFF (Summer, 1995)

The year is 1995. I am off on a camping trip and perched on the edge of a giant cliff above the St. Croix River in Taylors Falls, Minnesota. I have been cliff-jumping my way up to higher spots at twenty feet, thirty feet, forty feet, and so on, and I am now seconds away from the biggest leap of my life a seventy-foot jump. Looking down at the dark river passing below under the shadow of the sun, I have no idea how deep the water is or even where the appropriate landing will be after the jump. I have never seen it done before and have only heard rumors that it can be done. I am filled with adrenaline and the sense that by doing this, I will be one step closer to infamy and have a leg up on all those who've ever doubted my abilities.

Leaning in, I steady my feet on the granite rock's edge, bend my knees low, and slowly look over the edge one more time to assess the leap. Below me is a V-like gap between two sheer granite rock walls that drop directly down into dark waters.

Behind me, some ten feet, stands my college roommate, Tonie, who asks, "Do you think you can make it?"

Tired of waiting and having reached my conclusion of the feasibility, I whimsically responded, "There's only one way to find out."

Then, I turn, thrust upward from my squat, and launch over the edge, achieving as much trajectory away from the face and begin plummeting toward the earth. This decision to jump, at this moment, was one in a long list of decisions that I had been making in a similar way for as long as I can remember. It was a pattern of choices built around a desire to dream big, but while my intended desires were pure, my execution was high risk. It lacked planning, had no exit strategy, and time and time again, fell short of the goal, coming painfully close to my physical, mental, and spiritual demise.

Each time, I would find myself in a spot in which I was not prepared to be. Each time, I was forced to make tough choices of going forward or returning from where I had come. Many times, I fell short of the goal or failed altogether. Over and over, I would take on fear and jump, depart friends on high ground, plummeting into the deep unknown below, splash down in new waters, forced to sink or swim, and return to dry land with a new life experience and lessons before starting again. It was a pattern that would lead me to some of the darkest places in my life, teach me some of the greatest lessons on living, and force me to rethink everything that I had known about myself since I was a child.

On this day, however, after plummeting for what seemed like forever over the dangerous seventy-foot drop, I burst through into the brisk waters below and was alive! Painful as it was, I had survived with only the hard sting of my feet from the hard water surface and the rush of the cold river up my backside like an enema from a fire hose. With time, I would learn a new way to look at each cliff and a new path toward launching off toward my dreams. For now, however, I was depleted of adrenaline from the thrill and needing to find my way out of the deep water and back on solid ground.

CHAPTER 1

THE OBSERVER, THE DARKNESS, AND MY CHILDHOOD HOME (1979)

A special memory in life is recalling the long stretch of fields and beautiful patches of woods just past our backyard at my rural countryside childhood home in Medina, Michigan. From the age of five until I was well into my late teens, I lived on the outskirts of that small town. South of Ann Arbor and north of the Michigan-Ohio border, the town I grew up in was called Medina; it was a speck on the map with a population of about three hundred.

I can still remember the first time I walked into our big country home in Medina, moving from a small town home in Morenci, where I was born; it was a bit of a shock. The Medina house did not have neighbors across the street or sidewalks out front like our previous home. It only had trees, fields, and open land for miles around. The two-story, 2,500-square-foot home, painted white with green trim, was perched on top of a small hill overlooking the church and town to the north a quarter mile. In all its glory, this would be my home for the next fourteen years and become the origin of some of my greatest childhood adventures.

Over the first weeks and months of our time there, I started exploring the boundaries of the land around me. With fresh summer grass under foot, I started first walking around the one-acre yard and barn that stood on the lot next to our house. From there, I could see a small grouping of trees on the hillside just out from the kitchen window where Mom could keep an eye on me. Racing over, bound only by my shoes against dirt, I found myself quickly grabbing onto the lowest fist-size branch of a small twenty-foot baby oak; from there, I worked to connect my way upward higher and higher. Always pushing the limits, I climbed until the thin base at the top of the tree started bending back toward the ground.

Day after day, I would push the boundaries of our new yard. I traversed from the large maple trees in the front over the rock-retaining wall to the basement-level garage and into the dried-up pond, around to the giant weeping willow, and through Mom's garden in the back. I jumped, rolled, and ran like a dot-to-dot scene in a Family Circus cartoon until I'd known every square inch of that property like the back of my hand.

I am not sure how much time passed in my exploration before going beyond into the foreign woods and fields behind our house, but I do remember the first time I went.

In the country, there weren't a lot of kids to befriend, so when you made a friend in the family, you were forced to share, for better or worse. My brother Kevin made his first friend—a very stocky young neighborhood boy named Chris. On one day after they'd grown tired of going between houses and torturing me, they decided to explore the edge of the woods past the back of our house a half mile through the field that seemed to stretch forever. Having never gone that far and curious to see what was back there, I tagged along as the third wheel and did my best to keep up as they moved quickly along the fence row at the edge of the field. Knowing the tendency of the two of them to team up, bully and beat me to a pulp, I kept a certain distance—close enough to feel safe but not too close to agitate. Walking, I could feel my rubber boots squishing into the shallow mud of the fields. Softened by the rains of fall, I worked to navigate to the firmer high grass of the fence row. Step-by-step, I could feel myself moving away from the safety of our home and closer and closer to the edge of the woods and the dark unknown that was just beyond.

Soon, we began the drop down a small rolling hill into a valley that fed into the hidden edge of the forest. At the base of the valley, we uncovered a spring that transformed into a creek leading into the forest. At its base, there was a small shallow pool with a sandy bank to play in. It was an enticing spot to stop as it was close enough to the edge of the forest to peer in, yet near the open field to have an escape.

Here, Chris and Kevin paused to play on the water between the mud and the sand. Pausing, I looked up and noticed a giant old-growth oak tree covered with bright green moss that almost seemed to glow from the moisture of the spring. For me, it was

CHAPTER 1

like seeing a dinosaur walk through the frame for the first time in the opening scenes of *Jurassic Park*. The massive trunk was bigger than life, and the scale was enormous. Bending my neck, I looked up to see a body that seemed to reach up into the sky. It felt like this thing was calling to me to come closer. Walking around the root structure, I noticed the thick brown bark of its exterior protecting its softer inner core. Years of harsh storms, strong winds, and cold winters had come and gone, and age had left its mark. Wisdom, in a metaphorical sense, seemed to ooze out of the soul of this grandfather of time, and my imagination of its story was running wild. Walking around to the opposite side, I was now hidden from the play of the other two boys. Here, I found a long arm-like branch reaching down toward the ground as if to invite me in. Seduced by the idea of climbing higher on this beautiful beast, I started my way by stepping onto a massive branch and bear-hugging my way up. Slowly, I nudged upward, each shuffle forward, inspiring the next.

Finally, after several minutes, I reached a joint between two sections of the tree, a perch high above the ground where I could sit and look down upon the earth. Now below, I could see my brother and his friend playing in the mud. From here, I was safe from their mischievous nature and could not easily be touched. In fact, they didn't even know where I was or that I was watching them. For the first time, I felt in control of my surroundings. Up here, I was both with them and separate from them. In a sense, I was omnipresent, both everywhere and nowhere at the same time. In this space, I could observe and seemingly understand them without the chaos of rubbing shoulders with them. It was a feeling of empowerment that I remember liking a lot.

Lifting my head from watching the games below, I could see a long way into the distance. Far ahead from here, I could see the forest from the trees. Beyond was a massive section of wood that I had not yet discovered. Overhead, there were passing clouds that seemed to know nothing of time and space. Between the white floating giants above, light leaked down, falling into enchanting places below, creating windows into the forest that felt like they were inviting me to explore.

Now, for the first time, I was learning of a world I had not known existed, and something inside me desired to know more. Then, like a lightning bolt from the sky, I snapped out of my daydream to the sound of faint voices laughing and running in the distance. It was my brother and his friend Chris. Turning around from my perch up high, I could see them running at full speed back up the hill toward the house, leaving me in their dust. Suddenly, the world pushed in on me, and the space I had enjoyed felt very different. I was now aware that I was alone and needed to resolve this problem immediately.

In an instant, the forest that had once seemed romantically enchanting felt dark and lurking with danger all around. As fear swept into the void of my dreams, I felt panic and knew that I needed to escape. Scrambling, I climbed down the tree and back onto solid land. Then, with my feet hitting the dirt, I began to sprint back toward home through the light of the open field. With my heart racing, I could feel the darkness of the woods chasing me. Slipping through the muddy field, exhaustion was setting in, and it was all I could do to make it to the property line of my own home, where I knew I would be safe. Running with all my might, I went as fast as I could, praying that I would get home before the darkness caught me. Finally, I made it and ran into the house to tell Mom.

ANXIETY, MOM, AND THE FIRST GRADE

I have to admit, as a young child, I was a bit of a momma's boy. One of the earliest memories that my mom often shares of me was the time when she left home to give birth to my sister. I was only three years old when she left for the hospital and had no idea why she'd gone or when she'd return. At home with my father, I missed her, and I hadn't seen her for what seemed like forever. Kids did not go to the hospital in those days. After being in labor for some time, she'd delivered my sister and was finally able to call the house to check in. When I finally heard her voice, I went ballistic. My dad, having spoken with her, decided to hand

CHAPTER 1

the phone over to me. No sooner did I get a hold of it and hear her voice that I started to scream, "*I want you to come home right now! I want you to come home right now! I want you to come home right now!*"

Over and over, I yelled this until my mom was in tears and my dad had to forcibly take the phone away from me. A few days later, she'd return with my little sister Marissa, a cute little girl who was very sweet, a little gullible, and always innocent. She would end up being one of my first adventure companions climbing trees, building forts, and playing together in those golden years, where imagination rules supreme and fun can be had with only a few good boxes and some tape.

I did not know it, but I was a bit of an ambivert—both introvert and extrovert—which is the classic actor/comedian type—the type that loves the stage and loves the cave but struggles with navigating the high stimulation and awkward conversations of small groups, classrooms, offices, or parties. Because of this, I did not connect with people easily, but once I did, as with my mother, I became very attached.

One thing that was clear, as a child, I felt deeply alone when separated from the people I knew and trusted. I liked my tribe and didn't like it when they were away. Being alone felt cold, harsh, and gave me anxiety. When I was shipped off to school by bus at the age of five, this emotion hit the ceiling. Being separated from my mom and my home, then placed in a classroom with new kids and an adult I didn't know, felt unnatural. I instantly felt lost and wanted to go home.

In a full panic, I began to break down and cry. To the teacher, I was inconsolable; she was at a loss and decided to call home. After speaking with my mom, who was thirty minutes away, they found my brother and decided to have me sit with him, hoping it would help. Seeing his face brought comfort. I stopped crying, but the attention had caused him embarrassment. As he pushed his pencil down on to his paper and began to work, I could see him hiding his head low, and I felt a sense that he was wishing I'd just go away. Finally, Mom showed up. Relieved, I returned home only to discover I was going to be shipped back the next day.

Those first days of school were hard for me, but in time, I found a sense of comfort and familiarity in the classroom and settled in. Over time, they realized that I had a reading and spelling deficit and wasn't keeping up with the pace. No one told me directly, but I suspected as much. While I struggled in academics, my intuition was very high, and while they didn't measure it, my EQ was off the charts. I could read the teacher's mind, see how she thought in her eyes, and feel the tone of the environment in her words. When the classroom broke into small groups, I knew I was in the slower group, and I didn't like it, but I didn't have a choice. Then, by the end of the second grade, they required me to stay late for tutoring, and instead of riding the school bus home with the rest of the class, my dad picked me up in his blue Ford pickup truck on his way home from work. When I got in, he'd slide a Little Debbie treat across the big bench seat, and I'd gobble it up. It was a sweet memory during my struggles with learning that I treasure to this day.

In those earlier years, I could feel that I wasn't completely fitting in anywhere. Something was different. I knew from how the world responded to me, but it wasn't obvious what it was. I just knew that there seemed to be a track that the majority of my peers were on, and I wasn't on it. To cope, I found solace in nature and the outdoors. After the discovery of the big oak tree under the spring on the edge of the woods, I faced my fears of the dark spaces in the forest and returned to that spot. In time, that grand oak became both a home base and a portal to a much deeper exploration into the forest beyond.

As tough as it was to make it through my first two years of school, I overcame separation anxiety and a general sense that I was not keeping up with the class. With the help of tutors, I made some headway and settled into my routine. Then, as I entered my third year of elementary school, something happened that would change my perspective on education and the institution of learning forever.

MY PARENTS, JESUS CHRIST, AND THE CLASS CLOWN (1970 – 1979)

At the age of nineteen, after graduating, my parents, who were high-school sweethearts, got married. They did, like many youth of the "baby boomer" generation in small-town America in the 1960s, having both grown up with struggle and dysfunction brought on by alcoholism, illicit affairs, and unresolved conflicts. They wanted to offer something more to their kids—a life more positive than the boozing, cheating, and mistreating that had haunted some of their past. In their search for a better way of life, they were invited to and attended a religious revival headed by an evangelist (Rex Humbard).

Rex had started his work as a Pentecostal preacher in Little Rock, Tennessee. During the sixties and seventies, he had taken to the road of evangelism and could be found in churches around the country leading revivals. One evening, when he traveled to Morenci, Michigan, my parents sat and listened to him preach the gospel. Hearing the promise of redemption and salvation through Jesus Christ, they were both so inspired that they went forward to the altar, got on their knees, and humbly committed to serving God. It was an event that changed the very foundation of their lives. From that day forward, they began to completely reorganize everything according to their new belief in God.

By the time their three kids were born, they had become deeply embedded in what they would call their "home church" about thirty minutes from our home in Hudson, Michigan. There, they met a man named Paul, a principal in a small Christian school named Trenton Hills in the next town over. At the end of a service, Paul pulled my parents aside and shared the value of a good Christian education for children. At that time, my father was a factory worker, and my mother spent days as a check-out clerk. For blue-collar people from working-class families, the cost of going to a private school would be a sacrifice, but my parents decided it was worth it in exchange for a good Christian education. Forgoing family vacations, new cars, and general materialism for years to come, we were heading off to a new school.

To help make the way, my mom took a job driving a bus for the school and dropped us off on the first day. Sitting down in my classroom, there was a chill from the new desk seat as I started my third-grade class. One benefit in my mind of the change to a new school was that it allowed me to leave behind the failures that my previous teachers had known and make a fresh start. With some effort, I could hide the fact that I was in the "slow group" and get back on track with the rest of the class. Alone in this new world, I knew I was the only one in the room who knew how "dumb" I was, and I would do everything in my power to keep it that way. So, to cover up my weaknesses and protect myself, I became funny. Learning quickly that if I got called on in class to answer a question that I didn't know about, I had only to crack a witty response that would get a laugh, distract the class, and break up the tension. And if I was asked to read aloud, I would turn the reading into a theatrical appearance, acting like I meant to mispronounce the words or make them sound funny rather than embarrassingly stumble and stutter over words while trying to read.

Near the end of my third-grade year, I was nominated "class clown." It was even noted on my report card, and while my parents wanted it to stop, to me, it felt like a compliment or badge of honor. Still, the reality of the situation was that I needed help. I needed to slow down, listen, and work harder. However, none of that made sense to me at the time; I just wanted to fit in. Ignoring the reality of the situation, I kept doing "funny" until finally funny caught up with me.

By the end of the third grade, I had gone from the laughable new guy to the most popular guy. My classmates loved me. I was fun to be around and was the guy who didn't care. However, what I hadn't realized was that my popularity had blinded me to the truth of what was happening. I had earned acceptance by dismissing the very reason we were at school. And while I was playing, what I didn't know was that behind closed doors, big decisions were being made for me over which I had no control.

Before I knew it, the third grade was coming to a close. As the hours had turned into days and the days into weeks, the school neared its year-end. Those last minutes in the classroom on the final day seemed to be the longest, slowest minutes of all. Sitting in your seat, squirming to get out, time itself seemed to break the laws of physics and change pace, but when the bell sounded out, breaking the air with its high-pitch ding, it all sped back up double time, making up the difference. With the day over, the class was abuzz with bags and books flying as everyone was on a full charge to the buses and rides home.

This year, however, my teacher, Mrs. Marwelle, who had written "class clown" on my report card, asked me to stay over. Surprised, I sat anxiously at my desk, wondering what I possibly could have done so wrong during the last day that warranted this delay in my escape. During the past year, I had been asked to stay back multiple times for multiple reasons, though I usually knew why I was in trouble. However, this time, as far as I knew, I hadn't done anything wrong.

Sitting there waiting, my mind was racing. While everyone was running for the bus, I was in my seat in an empty room. Then, my teacher spoke in a low, slow tone so that I could understand each word coming from her mouth.

"Barry," she said, "before leaving, I wanted to share something with you."

Holding my breath, my heart was pounding, and I was still at a loss. My powerful skills of intuition had no read on the situation, and I just wanted to get this over and go home.

"I wanted to let you know that your parents and I talked," she continued, "and next year, you're going to repeat the third grade. We're going to hold you back."

Time, again, came to a screeching halt, and with it, my heart sank. She went on talking, offering intelligent reasons why, but I no longer was listening. I could only hear the powerful voice in my head trying to understand. Emotion was flooding into my brain, adrenaline was racing through my blood, and for the first time in my life, a hot red flush of shame engulfed my entire person. I could literally feel my face change color, and my body grow numb.

Finished, I mustered up the courage to grab my bag and walk to the bus. I greeted my mom as the manual double doors closed behind me. Pushing on the gas, I felt the acceleration as I tucked away inside a big bus seat, sunk into myself, and sulked as we headed home.

Inside, I wanted to help escape the shame of my glaring failure, but it wouldn't go away. When we finally returned to the country, I ran into nature to breathe and think. For the rest of the summer, I pushed farther and deeper into the woods, into the dark, unknown places of the forest, to sanctuaries where nobody was around to grade me on my performance and where I was safe and could feel solace. For the years that followed this event, I would be haunted over and over by low grades, failed tests, and incomplete assignments, all triggering the shame of being held back, not keeping up, and failure. This emotion would send me on a course of escape into places where I felt understood and ultimately far from the traditional path of my peers. For now, however, I just felt lost, and I didn't want the summer to end. I dreaded the day I would have to return from my escape to an institution of tests, assessments, and shame. I wished I could somehow change reality, become someone different, and escape from this place, but that day never came.

Barry in his back yard at the top of a tree with local friends hanging out below (shot from mom's kitchen)

Nathan's Volkswagen Beetle after the crash rolling it multiple times (1988)

"Wisdom is earned at the cost of suffering over time."

CHAPTER 2

AN INTRODUCTION TO GOD

In life, there are many events outside of our control, events that happen, both good and bad, in an instant. Those events end up shaping our perspective of the world. They help develop a narrative or story of who we are. The story often becomes a picture of how we perceive ourselves. It includes that which we're "good" or "not good" at doing. Over time, with the help of others, we often forget that there ever was a story and, instead, assume that who we are is fixed. We begin to live out on auto-pilot what is a fluid and changing story as if it were repetitious and predictable. This process can define who we are for a lifetime and, for some, can be a prison from which they never escape.

It was in the third grade that I had the first event that inspired a narrative of failure and formed a story of who I was. This narrative inspired a victim mentality. Supported by sympathizers among my well-meaning parents and friends, the mentality became solidified and reinforced by similar experiences that echo the first. This all contributed to what is known as a feedback loop that was nearly impossible to escape.

Well beyond the trauma of my elementary years, the story of being a victim of things beyond my control became hard to escape and easy to fall back on. Instead of fighting through to the other side and finding success, I leaned on my crutches and turned back early, time and time again, giving up on a new path out. This cycle went on until, one day, my world got turned upside down, breaking the pattern and sending me on a new search for something greater than myself.

GRANDPARENTS AND FAMILY ON THE DAVIS FARM (1974 – 1986)

As a kid growing up, I acutely remember the close-knit kin-like culture on my mom's side of the family. She grew up on a farm without a lot of resources and spent her life watching her father work the land and milk cows. That forced the family to work together to survive, and because of that, they grew up to depend on each other and understood the value of family. After marrying, Mom relocated seven miles away to the village of Medina and became the only child of five to move off the country dirt road known as Mulberry Road, where her four other siblings and parents lived out the entirety of their lives.

One of the best things I experienced as a kid was driving inside Dad's Ford pickup truck listening to the crackle of the family CB radio and the chatter from Grandpa coming in as we drove within range of visiting the farm. The family CB had been mounted under the center console by the advice of my grandfather. And as we hit Mulberry Road and came within range of transmissions, the fun began.

On any random night, as we drove up to the farm, you might hear the call sign for Dad. Grandpa would call out, "Chicken George, do you have your ears on . . . ?"

Dad would respond with a smile and a laugh, and over the next few minutes, everyone would join in talking back and forth and making jokes. It was a small, unique piece of our family culture that made us who we are.

Grandpa also loved fishing and the outdoors; he'd often go away on trips up north. When returning with a big catch from out on the Great Lakes, we'd get a call telling us to come over. We knew this meant he was going to cook up a fish fry for everyone to enjoy. On these evenings, the five siblings with their spouses, my aunts and uncles, would meet back at the family farm. There, under the mist of evening, all the grandkids would play and run around the front lawn. Hide and seek, kick the can, or some game of toss was going on while under a single light in the small old milk parlor, you could hear the sharpening of a filet knife as Grandpa prepped salmon, brown trout, or northern pike to be eaten. With a bowl of twenty or more filets each getting dropped in batter, you could hear the sizzle and smell the fresh cooking as Grandma would go to work with the girls in the kitchen, preparing food for the whole family to eat.

Those nights were rich. The experience of closeness and connection that passed among us had a texture to it. We belonged because we were kin, and it dripped off us like sweat. We possessed ownership of something that couldn't be taken away. You could feel that you were part of something bigger than yourself—something that possessed real-life struggles, hardships, good times, and bad with all its imperfections. It was a special feeling that I cherished and one that I would later spend much of my adult life in search of regaining.

The romance of those nights was only part of what made us family. There was also history—moments that went unspoken but lay just beneath the surface. If you scratched just enough, you'd uncover untold stories that impacted us in ways that would go unshared for a lifetime.

As a young adult, I would visit Grandma's house on the farm long after the fish fry era of my childhood had come and gone. By then, Grandpa had passed away from lung cancer, and Grandma sat alone in the quiet of her home. Walking in, I would sit down next to her on the smallest of couches. Looking closely, you could see that the wood and varnish of her favorite chair had been worn down from all the hours sitting and rocking while staring out the window. Rocking ever so slightly, she would turn to look out the sliding glass door toward the cows in the field when she was thinking. A small woman, only 4'6" tall, I looked like a giant standing next to a hobbit when I would stand up to take a photo or say goodbye.

When visiting her, we'd always have something simple to eat and finish with her special homemade cookies that she'd pull from the freezer. By this point in life, the grandkids were all grown, and time had moved far beyond the simplicity and fun of those CB radio years. Recovering from lunch, we'd rest quietly enough to hear the old clock ticking on the wall to the sound of "click-clock, click-clock." It added to the ambience of the place.

Often, in the passing of time, Grandma would speak up and offer some life experience and wisdom amid stories of working on the farm and missing her kids. She'd often just blurt out phrases that would make you think. Of all the phrases that stuck in my head over time was the one that was near biblical in origin. "Barry, remember one thing," she said. "If you screw up, you'll reap what you sow, so mind your business." Then, she would smile. I knew in that smile was a lot of love.

I loved her dearly and idolized her, but little did I know that there was much more to Grandma's statement than meets the eye.

Grandma used to tell me how on a sunny day when my mom was just a baby, she'd wrap her up in a blanket when the hay needed to be cut. Then, she would set her in the bassinet on the side of the field and get busy on the tractor cutting hay. It's hard to imagine that little baby laying there vulnerable to almost any danger, but that's just how things were done back then. There was no one else to help, so Grandma prayed the baby would be safe and got to work.

By the time Mom had become a teen on the Davis Farm, things were more established. Being the youngest of five children, the other four had grown. Some had married, but the ones who weren't had chores to tackle. Ivan, her brother, who was set to inherit the farm, was now doing the majority of the bigger tasks in the field. All the things that Grandma once had to do to survive were being done by others on the farm. Unbeknownst to her, she had become a mid-forties country housewife. And for the first time, she started to have free time, with no idea what to do with it.

CHAPTER 2

While Grandma was home during those years, I assumed Grandpa was still very busy and seldom around the house. In the spring, summer, and fall, fields needed planting or sowing. When he wasn't out on a tractor doing that, the cows and animals on the farm needed tending to. When he did have free time, Grandpa would usually be out doing what he loved, hunting or fishing. There were no thoughts of things like date nights, which were not really holidays observed on the farm. Knowing now what I know about life, I imagine that the romance of marriage had begun to run thin, and Grandma began to feel the sting of midlife.

The click of the clock on the wall that ticked in the silence of that house now sounded different for Grandma at age forty. Time was passing her by, and inside, I imagine, she was struggling to feel alive. I guess it was in that space that she was vulnerable to temptation and it's in that space that mistakes happened, which she would regret for a lifetime.

While the details of all that had unfolded are gray, I know it was the neighbor man in the rental home, and I know that it did not go well for him once Grandpa found out. What I don't know are the details of how it happened. I assume, like any affair, it began with a friendly conversation out on the lawn, possibly getting the mail from the mailbox on the road's edge. The day was grinding by slow, and Grandma just enjoyed the attention of feeling attractive again. Whatever the case, after some time, Grandma fell prey to a romance outside of marriage. And like so many bad things that happen in life, it was too late when Grandpa finally figured it all out.

As it was told to me, Grandpa nearly killed the man with a shotgun and would have done so if it hadn't been for his son Ivan's intervention. In an instant, shame fell upon the house. The sunshine on the bassinet in the hayfield turned to gray skies. The foundation of my mom's life had been broken, and everything began to change. That moment would define every decision, directly or indirectly, in the family for ages to come— one that would impact my life without my ever knowing it existed.

Long after Grandma had passed away, I learned the truth about the affair and understood her warning that day, "reaping what you sow." It was an important piece of history, one that helped me make sense of who we were as a family—a family that went to church every Sunday, come hell or high water.

THE ROAD HOME AND THE CAR WRECK

Church, like life, is full of a mish-mash of different members of the local community coming together with a common goal to worship something greater than any one individual. The members aren't perfect; in fact, they are full of imperfection. But they go there to pay homage and strive for a better way to live life. This still resonates with me as a very valuable practice today.

As a kid, I didn't really connect with that part of the experience. I went to church mainly because I had to. However, in time, my friends became a big motivator for going. It was also at church, at the age of nine, that I met my best childhood friend Nathan. Upon our introduction, I instantly liked his rebel spirit. He liked to blow things up, enjoyed being mischievous, and wasn't afraid to be himself. For me, that was a strange combination, which was very attractive as a boy.

Nathan also had a similar bent for adventure that I did. One time, we discovered an old rickety, broken-down bridge that had primarily collapsed and was no longer drivable. To him, this was a rare risk that could not be missed, and he led the way, charging across its old beams. Unable to resist a good challenge, I followed. On another occasion, there was an old, abandoned barn that should have been condemned. Charging onto the rotten floor, he climbed up into the hay mow twenty feet, then convinced me to enter and climb to the rafters even higher. I am still filled with fear at the idea of how that thing could have collapsed at any minute, swallowing us up with it, but we went on to explore it like it was a Roman ruin. Still, there was another time when we found some water run-off pipes four feet in diameter, jutting out of a reservoir dam that held the big lake in place.

CHAPTER 2

Those pipes ran fifty yards underground into pure darkness and ended at the basin of the dam's drainage. Looking inside it from one end toward the other, you could only see a speck of light encouraging you to enter. Taking the first steps past the outside, we went into the darkness with snakes and spiders. As we cried out whoops of fear and excitement, you could hear the echo of our voices travel from end to end. Within a few minutes, we were deep inside the belly of that pipe and nearly to the other side.

The crazier, the scarier, the more out of control it was, you could bet Nathan was leading the charge, and from the age of eight till the age of twelve, I always followed shortly behind him. As Nathan grew older, the risks he took grew bigger and more dangerous. What was once fun become just plain crazy. Then, on a snow-filled winter day, he went too far.

Covered deep in snow, the hills behind my house were magical to look at after a big Michigan winter storm and more fun to play. We'd bundle up and escape the warmth of the home and go out into what literally seemed like a wonderland. For hours, we'd roll snowmen, slide down hills, and make snow angels, all before going back inside to a warm cup of Mom's cocoa with marshmallows. After one winter storm while outside playing, Nathan showed up in glorious fashion. His dad was out of town, and since he was gone, Nathan had snuck away with his Ski-doo Formula MX with 462 CCs of power. When riding on the back, you could feel the pull of this thing as it instantly jumped up from zero to speeds of up to 75 mph, insane for a child of Nathan's age and capacity for adrenaline-racing thrill.

Grabbing me to go with him, I straddled the long bench seat and wrapped my arms around his waist as we shot off into the woods, racing through trees and trails. Then, breaking out into a clearing, he spotted something. There, in the middle of a field, he came to a stop and pointed. In front of him was an old farmer's lane with a long, steep ditch rising out of its side that took the perfect shape of a ramp. Looking at it, I knew he had an idea, but I did not have time to offer input or get off the sled. Suddenly, he hit the throttle, jumped up to speeds over 40 mph, and said, "Hang on!"

In an instant, we were racing the football field distance toward the ramp with no holds barred. Cold wind blasted my face, and my eyes lids were flapping open and shut. Then, without hesitation, he raced forward and hit the ramp, sending us flying into the air. Below us some twenty feet was the snow-covered road. Suspended in air, time had slowed, and it felt as if we were a fixture floating in a snow globe. Having traveled some thirty feet or more through the air, the vehicle came crashing down, and Nathan landed with it, but I was not on the back. The process of hitting the ramp on the long bench seat functioned like a catapult for me hanging on the back, and I had been sent on a slightly higher and longer trajectory. Having watched him land below and jet away, I dropped hard on my side into the snow and rolled over flat on my back. In pain and covered in snow, I looked up at the crisp blue sky and had a moment of clarity. I thought that if I kept hanging out with this guy, he would kill me. Shortly after, I stopped, and our friendship began to fade, but like any bad addiction, I went back one more time, and this time, the devil went back with me.

THE RIDE TO THE OTHER SIDE

Three years had passed since the snowmobile event. I was fifteen, and Nathan and I had all but stopped hanging out. If I did see him, it was at church and usually only in passing. Being a year older than me, he was now old enough to drive and had secured a license and car. For his birthday, he had saved up money and bought a used red Volkswagen Beetle (aka VW Bug) and could be seen racing around the town with his usual reckless abandon. My parents, knowing the dangers of such things, forbade me to ever ride in the car with Nathan, which was pretty much fine by me since we didn't really hang out a ton anymore.

Not having Nathan as a friend to bum around with, however, made church all the more difficult to bear. During my preteens, attending church had been about friends, but by the time I reached my teens, it wasn't. Most of the friends now were from my school or in sports. Without them, church had just become a mandatory and painfully boring experience. I just wanted to get in and get out, but I still had to attend weekly, and door-to-door, I had the schedule down to a science.

After Mom would painfully walk through the house shaking and waking everyone up and getting us into the car, we hit the road and pulled in around 10 a.m. Socializing in the atrium went till 10:30 a.m., which I didn't mind too much except for the occasional ear grab by an elderly member who hadn't seen me for some time. Then, we'd file into our seats and sing, which felt like an exercise of standing and sitting to wake up. At 11 a.m., with the audience primed to listen, our pastor was invited to the stage to preach. This was the most painful part of any service. I was a hyper-squirmy kid, and doing nothing but listening to this long, drawn-out speech was like sitting on a bed of pins. It was all I could do to keep from blurting out, and with some luck, I'd find a pen and pad and start to doodle. If we were on time, our pastor would end his preaching at around 11:45 a.m. in time for one last song. Then, dismissal; it was here that I was finally thanking God because, with a little fanfare, we were heading for home.

There was an exception to all this, and that was the advent of Potluck Sunday. On these Sundays, people would walk in the church early with big bowls of salad, pans of lasagna, or plates of whatever they'd made at home. When this happened, my personal schedule was shot to hell. After the service wrapped up, everyone would head down to the fellowship hall to eat, talk, and do what adults do for an indefinite period of time. I admit I did like the food options, at times, and enjoyed walking around saying hi to my favorite people like good old Tim Guest or Paul Palpant. However, once that was over, I was stuck—waiting till Mom and Dad finished.

It was on Potluck Sundays, where, to get home early, I would break down and beg my older brother for a ride. He was seventeen and had his own car. It came in very handy as an escape vehicle to get home early. However, he seldom wanted me around and preferred to ditch me versus help me out. The experience was always unnerving. But the trick was to first convince him to let me ride. If he agreed, I'd have to pay close attention since he'd never tell me when he was leaving. Then, when he moved, I moved, and once outside next to the car, he couldn't say no.

On this particular Sunday potluck, the day my life would change forever, I had been distracted talking, and when I picked my head up, my brother had already gone out of the Fellowship Hall. Scanning the room, I noticed his absence and raced to the parking lot to catch the car. Running out onto the dark black top, I could see the empty space against the yellow lines where his Toyota Corolla was once parked, and I let out a sigh. He'd made it away, and I was still stuck here waiting.

Disappointed, I picked up my head from the empty spot and noticed the red Volkswagen Bug that Nathan owned still parked. Right then and there, my mother's words rang in my head, "You are to never ride in that car with that boy." Quicker than lickety-split, the rebel came rushing to the surface, and I had an idea that I needed to act fast upon.

Racing into the church, I went from the sanctuary to the main hall, then to the bathrooms, with no luck. I was looking for Nathan, and I knew that if I could get him to leave early without telling my parents, they would assume I was with Kevin and would never know that I had broken the rules to get a ride home. Finally pushing open the back doors of the church, I found him standing alone, smoking a cigarette. Without haste, I shared my plan of hitching a ride home in his car to sneak out of church. Always up for some mischief, he agreed, and the two of us immediately slipped out the side door, jumped into the VW, and headed down the road.

There are a few details that you should know about the road home from church. From end to end, it was paved and well-managed. But if you don't know what you're doing, there are some risky sections. Upon leaving the church parking lot, you turn south onto Highway 127, heading toward Ohio, and drive for about ten miles. This section of road was straight, flat, and often patrolled by state police because of how easy it was to speed. After driving for approximately ten minutes, you reached Medina Road, the road home, and turn west for about fifteen miles.

CHAPTER 2

In a flat land of farms and fields, Medina Road had a rare hodgepodge of rolling hills with three very dangerous curves. These three curves twisted around the edge of Potters Corner and sat above a hill where, in the winter, kids would come to ride sleds as it was the only steep embankment for miles around. For anyone driving safely, this section was not of concern, but for a teenager like Nathan, this was the highway to the danger zone, and I had just jumped into the aircraft about to take off.

Sneaking back out into the parking lot, I almost felt like a bank robber on a getaway. Looking around, I could see that no one was watching as Mom and Dad were back in the church busy washing dishes and chatting with friends. Although the bucket seats in this old car seemed to be held together by glue and the old beast raddled like an iron cage, I was happy as a pig in shit to be on the road and out of sight. Thrilled by my savvy escape plan's success, I felt confident, and as we turned onto Medina Road, I started to relax. Looking over at my getaway companion, I began to remember the good old days with Nathan and questioned my judgment. I wondered if I'd been a little hard on him. In a way, we had all been a little hard on Nathan. Possibly, he wasn't as unstable and unpredictable as everyone had assumed but was actually quite responsible. While he was driving with two hands on the wheel, I looked over at him smiling and said, "How fast does this car go?"

Instantly, I knew it was a horrible question to ask because no sooner had the words come out of my mouth than the devil himself showed up. I had only asked him in that moment looking for a verbal response. I would have been happy to hear the words 70 mph, but for Nathan, it was a test of his manhood, and he instantly pressed his foot down to the floor and declared, "Let's find out."

It's funny how the human mind works. You escape church to race home without your parent's permission with a person with whom they expressly forbade you to ride. And the whole time you are doing it, you're nervous and looking over your shoulder, hoping not to get caught. And then, when you get just far enough down the road to think you've gotten away with it, life teaches you something important. It teaches you that no one gets away with anything, ever.

As I was watching the speedometer grow in this glorified go-cart, I knew at some level that I was in trouble. As we sped up to 70 mph, the car began to shake far more violently. Its old frame was warbling and rattling under four poorly balanced balls of rubber cascading down hard pavement. Dust, scraps, and shavings of paint were flying off the back, and by this time, I had seen enough to know that we had reached maximum velocity, and I wanted to slow down.

Cresting the final of a group of hills before Potters Corner, I knew we were charging full speed toward some trouble Nathan knew nothing because the ride to his home was different from mine, and this section of road wasn't one he often took, or ever for that matter. In my mind, I knew I had only a few minutes to educate him on the dangers. But he was now in a world of his own. And it seemed the more I asked him to slow down, the more it encouraged him to race ahead.

With his foot still hard on the pedal, he was about thirty seconds away from the big three corners of death. Feeling the danger ahead, I decided to come out of myself and be more direct. "Hey!" I yelled over the roar of the engine and rattling car. "We tested the car enough; we can slow down now, right?"

Turning, he smiled with the face of a devil and didn't respond or reduce the speed. Realizing we were in trouble, I started looking for ways to weather the storm. I reached for my seatbelt, but it'd been cut out of the car. This was a common thing to do in the eighties when seatbelt laws were just beginning to be introduced, which seemed insanely stupid at the moment. With no seat belt and only seconds away from the turns, I was enraged with fear and began yelling for Nathan to slow down. Intensely focused with his eyes on the curves, he could not hear.

With the road racing underneath rocks kicking up and trees flying by, I braced myself for the worst while Nathan's mind mapped out the course. I knew that for Nathan to make this, he would have to make three turns. The first was a slight right turn at fifteen degrees. While the caution sign read 30 mph, he could easily take the first curve at the speed we were going. However, the second curve was progressively steeper and more dangerous. For this one, he would need to turn left at some 45 degrees and slow to under 55 mph; if not, he'd be forced to cross into oncoming traffic around a blind turn. This was the most dangerous maneuver of all because the trees and foliage on the driver's side blocked the view. Therefore, to make it, when you pulled across the double yellow line and into opposing traffic, it was a gamble, and all I could do was focus on the road ahead and hope for the best. If no one was coming on that blind turn and fortune was on our side, we'd possibly make the third curve an easy turn and escape unscathed.

    Hurtling forward, there was no more time for bargaining. Finding the "oh shit" handle above my head on the passenger door, I grabbed on for dear life and pressed my feet hard into the floorboard, driving my back into the bucket seat and locking myself into position. With muscles tense, we raced into the first turn, and I could feel the weight of the car shift over the old springs and donut tires. Passing the apex, we started toward the second curve, and the car began shifting its weight back to the other side. Crossing the double yellow line, I began praying harder than at any time in my life, hoping God could hear. This was the moment of truth driving against traffic, and truth showed up in the form of an oncoming car.. Coming from the opposite direction and traveling at full speed, all I could see was a grill, a windshield, and wheels. Within fractions of a second, Nathan turned hard back across the double lines into our lane, avoiding a head-on crash. At a glance, I saw a green sedan swish just past our door out of sight, never to be seen again. All of the momentum and weight of the VW Bug that had been leaning to the passenger's side now shifted hard onto the driver's side of the vehicle, compressing the springs and shocks to their maximum capacity. The cheap rubber wheels were doing everything they could to hold onto the road as our direction was now charging directly toward the ditch ahead.

    In a last-ditch effort to save the ride, Nathan turned hard back into the curve in hopes of somehow recovering the turn.

This final move threw all the weight of the car back to the passenger side, sending us into a roll and sealing our fate. To my right, I could see the ground come crashing toward me into the window. Gravel, dirt, and glass sprayed into the passenger seat. Around me, the world was spinning out of control; I was now losing my grip and could not steady myself against any one part of the car. Suddenly, in a full barrel roll down the road, my world went black, and the ride soon came to an end.

Opening my eyes, I was trying to make sense of where I was. I felt like I had just woken from a dream and had appeared here out of thin air. I was dressed in my Sunday best, but my shoe was not on my foot. The smell of gas was pungent in the air, and I couldn't make sense that I was in a car, and the windshield was gone. Not certain where I was or how I'd gotten here, I knew I needed to get out, and I struggled to exit the vehicle. Crawling on my hands and knees, I made it to the road but could not stand up.

Turning back, I could see the full scale of the damage. The roof was concaved, the windows were all shattered, fenders and hub caps were strewn across the landscape, and the car axles were exposed to the sky with one wheel continuing to rotate in place. Slowly, my memory returned, and I recalled I had not been alone. Still on my hands and knees, I began to look for Nathan. He was not in the car and nowhere to be found. Calling out, I began to look in all directions, perplexed by where he could have gone. Then, in the distance, I could see his body face down on the side of the road. He wasn't moving or responding. Crawling over, I reached out to shake him, and then for the first time since the world had stopped spinning, he made a groan.

Exhausted and in shock, I rolled up next to him and laid down. Now, the adrenaline was fading away, and the pain of all that had happened to my body began to creep into every muscle. My internal organs now felt like mush, and my back was killing me. Peering up at the blue sky above, I began to feel myself slipping away. It seemed like death was not as far as I had once thought, and my mortality suddenly hit me. Looking up at the vastness of space, I called out to God.

"God," I cried. "I don't want to die . . ."

CHAPTER 2

"Please, God," I begged. "I don't want to die!"

In a panic, I cried again, "God, if you heal me, I will do your work for the rest of my life."

Just then, I felt a shift in reality. In the distance, I could hear people coming. Sirens echoed from afar. Help was on the way. In an instant, my world had been turned upside down. And I would never see it the same way again.

Cliff Jumping in Taylors Falls, Minnesota (1995). Barry launching a 25-foot cliff on his way up to the 70-foot cliff. College roommate Tonie looking on.

"The gates of hell are locked from the inside."

—C. S. Lewis

CHAPTER 3

BIBLE COLLEGE, LOVE, AND A SIMPLE PLAN (1994 - 1998)

THE UNKNOWN TERMS OF GOD

I have often lived with the belief that people give advice to others that they first give to themselves. One might consider when taking advice, if you are not careful, you may end up living a life that isn't yours.

Three years had passed since the car crash. The sounds, the pain, and the trauma of that day had all but left; however, that moment of looking up at the heavens and crying out to God still lingered in my mind. Having walked away from that deadly accident completely unscathed, I was convinced that a miracle of some sort had occurred. I have known kids who have been in lesser or equal situations and have not survived. A big part of me felt like I had been spared for a reason. In the heat of the moment, out of personal desperation and fear of death, I had bargained with God, promising, in exchange for my life, to serve Him and do his will. In my mind, there was a sense that I owed Him something, but it was unclear exactly what that was.

Within this was a new chapter in the story of who I was becoming. In my personal narrative of self, I had told myself that I was a failure, but now I was a failure who was saved by God. That redemption by a higher being offered salvation and the hope that my life had a purpose and was destined for something great. While highly simplified, it was this practical belief that paralleled the Christian faith that began to propel me forward.

At the end of my senior year of high school, I was a little lost on what to do and did not have a lot of direction as to what was coming next. I had not been a great student, my grades had been historically below average, and I was arguably not on the college track.

I was battling this internal turmoil about what to do with my life; my parents were insistent that I needed to go to college and get out of the house. Wanting to please my family and avoid joining the workforce straight after graduation, I started to look for colleges for which to apply.

In my search, I befriended an eighty-year-old minister from church named Pastor Skoog. Skoog, as we called him, was a mentor of sorts. He had spent his life working as a minister and traveled extensively throughout the country doing church planting. At the end of my senior year, I traveled with him by car down to and through Missouri to visit some of the old churches he'd started. It was one of my first big road trips out of the state of Michigan.

Driving through the Ozarks was my first introduction to the Crystal Rivers and deep rolling backwoods of this part of the country. I was entranced watching from inside the car while Skoog shared of his life of forty or more years past. Over the miles of road, he told me how miracles had occurred and how God had provided both financial and physical needs in times when there weren't any. During our time out there, we'd visited small churches with congregations of joyous elderly people who remembered Skoog from years past. In these spaces, I'd seen the fruits of his labor from sharing God's love and building communities. The message he shared had spread to hundreds if not thousands of people changing their lives and their children's lives for the better. It was a formative journey that opened my eyes to the value of sharing the love of God with people as a pastor and leaving a place better than how you found it.

After our trip, I contemplated my bargain with God and began to question how I would fulfill my obligation. In the most literal sense for me, becoming a minister was the most practical application I knew to fulfill my promise to serve God. As a course of action, I started applying to Bible college. After working to navigate admissions and get my grades high enough to be accepted, I was accepted as a student at North Central Bible College in Minneapolis, Minnesota. In the winter of 1994, I arrived for freshmen orientation and had no idea of what I had signed up for or what I was going to learn.

Prior to this, I had felt that college was never really in the cards. In fact, after the car incident, I decided to join the Army National Guard, thinking this could be a career path. In 1992, I signed a six-year contract and went off to basic training in Fort Jackson, South Carolina, and completed my Advanced Individual Training (AIT) as an electrician at Fort Leonard Wood, Missouri. I finished the program at the top of my class and on course to enter the trade upon returning home. But the reality was that I wanted something different for myself. I believed in dreaming big and living big, and kicking off a career to be an electrician in the small town where I grew up didn't feel like it was that.

Alongside my conviction to God, I also wanted to perform on a stage; that class clown from the third grade still wanted to be the center of attention. However, the biggest stage in my small-town view of the world was behind the pulpit, speaking. As misguided as a motivator that was at that time, it was the attention of being center stage that ultimately tipped the scale and drove the final decision.

The experience of Bible college could be captured in a film of its own making. From the get-go, it felt like a mix of the respective films, *Dead Poets Society* meets *American Pie Presents: Band Camp*. In other words, it was like a watered-down version of Harvard meets the dorky zit-faced kid who doesn't get the joke. In short order, I discovered that the place was full of stifling biblical courses taught by scholars who loved biblical history and the legalism of religious rules more than the practical application—a major snooze factory for me right out of the gate. Those scholarly faculty members were juxtaposed by a vast array of sheltered pastors' kids who were often torn between the reckless ambition of their youth and a desire to live up to the expectations of the institution of the church. Most of the boys were there because their dads were pastors, their moms were pastors' wives, and they were following in their family's footsteps versus seeking a deeper understanding of God and the practical application of his Word. On a similar note, the girls were mostly there to meet a future pastor and become a pastor's wife. All I knew was that I was there trying to live out a promise I'd made to God and figure out who God was in the process.

During orientation, I was informed that chapel was mandatory five days a week. Every day at 10 a.m., everyone would pile into the auditorium to sing songs, listen to sermons, and pray. It was a ritual designed to tap into the Spirit of God that sometimes worked and sometimes just felt redundant. Afterward, we'd all head to class. As a freshman, I was taking a hearty curriculum of Bible courses that were mostly over my head. The reading was miles deep, and the topics put me to sleep on page one.

It is worth noting that as a result of my failures in the third grade, I learned as a survival technique to cheat and cut corners on everything that didn't come naturally to me in academics. From essays to timed tests, I had become an expert on cheating. I am not proud of it, but I can say confidently that up to this point, I had never read a book cover to cover. While I had signs of dyslexia and ADHD, these were never diagnosed, and I only knew it was hard to get past page one. To survive, I learned to skim or buy Cliffs Notes and flub the answers on the test.

When I entered college, I was committed to doing it differently. I wanted to go by the book and prove to myself that I could do it—like in the movie, *Butch Cassidy and the Sundance Kid*, when the two main characters moved to Bolivia and decided to leave the days of robbing banks behind—I, too, had decided to go straight. Like them, I did my due diligence at least for the first few months. At first, my days and nights were full of schoolwork with hours at a desk turning pages. However, no matter how hard I tried, it didn't get any easier to study. My mind often drifted off topic, and most of the time, I didn't remember anything that I read.

With all the stress of trying to go straight, I needed an outlet. Living here in the middle of the "big city" of Minneapolis, where the college was located, I did not have easy access to nature, but I did have a new type of metropolitan escape—skywalks.

THE UNNATURAL NATURE OF THE CITY

While Michigan is cold, winter in Minnesota is next-level cold. Jet streams from Alaska dip directly down in a big U shape over Minnesota and bring that arctic cold down to the city of Minneapolis, freezing it to the bone. In 1996, I got a true taste of winter as a cold front in February was so bad that we went five days at twenty below zero and reached a record cold of -60 degrees Fahrenheit. Schools were canceled to avoid children getting frostbite from just going outside for only a few minutes. On the way to school during the freeze, the door handle on my Volkswagen literally iced shut.

To be able to work in the midst of this cold, the city devised a network of closed walkways called skywalks. These pathways traversed over the streets and connected to a network of skyscrapers and buildings throughout. Once you entered the skywalks, you were connected inside end-to-end for nine and a half miles, eighty full city blocks, making it the largest skyway system in the world. Literally, you could go to one part of the city on foot and not exit outdoors for hours.

The best part of this network was something I'd never found in nature or the small town where I'd grown up; it was a carnival of stores, shops, and experiences. Passing through one building, you might smell all the flavors of your favorite sandwich shop or, in another, popcorn popping. Exiting a building, there might be a long list of candy shops with the smell of taffy in the air. Across another sky bridge, you might find a grouping of apparel shops or catch the smell of leather from a Coach or Gucci store out of which women wandered. It was an amazing maze of wonder, and I loved it.

The shops and food were only part of what I loved. Some of the nooks and crannies led to elevator shafts, and those elevators went to the top floors of buildings with windows and views looking out over a new section of the city where you could cast your eyes to the horizon. Out there during winter, massive lakes were frozen over, trees looked like hordes of dead broccoli, and in the spring and summer, the little dots of people could be seen clambering around like ants below. Sitting up there looking down felt like an extension of my childhood fascination with climbing trees, and the observer in me found solace there.

I remember the first time I stumbled from the skywalks into "Gaviidae Common," an elite shopping center in downtown Minneapolis. Entering, I crossed the marble floors to look up and see the six-story open atrium that seemed to reach to the sky right there in the center of all those miles of enclosed glass. Looking up, you could see the natural light casting long lines across the open space from small windows high overhead. Sounds of the cascading water falling from the golden lip of a fountain forty feet up were trickling from the sky into a pool below. Stepping over the gap, I began to travel up the escalators to the top where you could look over the entirety of the space. High above the ground, I sat down on a bench surrounded by white marble floors. Observing the origin of the fountain high above, I sat beneath a celestial atrium and began to think.

Here, amid this unnatural beauty, I was overwhelmed with awe. It felt like the space was a place of worship. The fabrication of nature seemed like its own form of religion seducing me into a "material world." It possessed nearly everything I loved about the outdoors, but it was missing a critical piece. There were no bugs, mud, dying trees, or cycle of life. It was as if I was living in a dream of what nature could be, a utopia of what one person believed it should be, but the vision was absent of reality. It was a small glimpse into the simple seduction of pretty ideas painted on the walls of an artificial world.

LOVE UNDECIDED, LOVE LOST

As winter turned to spring in Minneapolis and flowers bloomed around the college campus, I met someone named Kellie and fell head over heels in love—a beautiful, fun-loving, and petite 5'3" blue-eyed blonde from Kansas City, Missouri. She was a pastor's daughter, feisty as could be, and was playing hard to get from the day we met. Her spunk and spirit sparked a red-hot passion within me, and in no time, we were off to the races.

On our first date, I took her back to Gaviidae Commons, the aforementioned beautiful mall in Minneapolis that felt like the high cathedral of materialism. Sitting close to her on the fifth-floor bench, I could feel her thigh push up against mine in sight of the golden waterfall. Filled with impulse and the raging hormones of youth, I reached out and grabbed her hand, resisting the urge to pull her in for a kiss. Instead, I recited the words from the *18th Sonnet*, a poem by William Shakespeare, "Shall I compare thee to a summers day, thou art more lovely and more temperate . . ." I went on to romance her.

Quickly skipping through skyways and elevators, I led her in a rush of excitement to a second favorite spot up a freight elevator to the fifty-first floor of the IDS tower. Here I had discovered an empty law office, and with doors open, you could go in, sit down, and overlook the entire city. Leaning into the thick, cold glass spanning the horizon, I could see more than just the physical elements in the distance; I could also see the metaphorical vision of the future. Lives together flashed in my mind, and I could no longer contain my excitement. Grabbing her close, I moved in for our first kiss and felt myself falling in love.

While the romance was high even after our city adventures, I could tell that Kellie was not as sold out as me. The more I moved forward, the more she pulled back, so I decided to double down. For the final months of school, I kept chasing her till the summer hit. At the end of the school year, she returned home to Kansas City, and I returned home to Michigan. When we left campus, I promised I would write her a physical letter every day. For the next two months at the end of every day, I'd pull out a pen and paper, scratch a journal of my day, then lick a stamp and throw it in the mailbox. Whether it was just before bedtime, after a long day of work, or on a weekend away with friends, I found time to write to her. On the other end of the post office, she would sometimes get five or six letters all at once. For me, it was a flood of love and passion until she finally gave in and decided to love me back.

By the time we started the second year of college, I had won her over, and it was official; we were together. This was very exciting—my first love and a first girlfriend. However, upon my return to school, she announced to me that she had opted to drop out of her studies and wanted to focus on us and work. This was a shock that shifted the entire dynamic of the relationship and forced me to question what I really wanted from both school and my new love.

Her absence from college meant she was looking to me to step up and provide. But I wanted her to figure out her career while I was figuring out mine. Now, she was playing the role of the pastor's wife, supporting the guy who was going to school to be a pastor, and the whole thing drew into the light the fact that I wasn't sure this was what I wanted to do. I was not prepared to be an ambassador of the religion, aside from the fact that I wasn't even a great member of the congregation. I had a lot of questions that were still unanswered, and seeing myself living out a life in the institution of a church and functioning as a pastor was in question.

The fact is that if I had gotten in front of a pulpit and was able to articulate what I thought, I would be considered a heretic at best and a false prophet at worst. My ideas began to come from outside the box of religion in nature. I began to think I saw God more in the place where I found solace than in the institutions that claimed to have captured his essence. I began to think that religion itself was a human attempt to understand God more than a definition of God's will. Its rituals created to help connect humanity with God felt valuable at times and rigid at others; however, the dirt path to the high trees in the midst of woods seemed to possess something that church could not and called out to my deeper sense. I felt connected there and could find application. It was a monotheistic view that reflected a mix of science and the biblical understanding from Psalm 19:1–4, "The heavens declare the glory of God; the skies proclaim the work of his hands. Day after day they pour forth speech; night after night they reveal knowledge. They have no speech, they use no words; no sound is heard from them. Yet their voice goes out into all the earth, their words to the ends of the world."

At the age of nineteen, if I had articulated these words in the halls of my Bible college, I would have been sent packing. While I was in love with a girl and seeking answers to her and the divine, I was not prepared to deal with the rejection of sharing what I really thought. As a result of following the path of a pastor, getting married to the pastor's wife gave me cold feet, and the more she wanted me to settle down, the more I pushed away.

Adding to the conflict, we had started having sex outside marriage, a sin in the Christian faith, and she was overcome with guilt. Looking for answers, she had turned to her parents for help. On a late night alone in my dorm, I received a call on our landline and recognized the caller ID as Kellie. Picking up, I was surprised to hear her father's voice, the pastor, on the other end. Having decided to help intervene, Kellie's parents had suggested a three-way call, and now we were all on it together.

Introducing himself to the conversation, Kellie's father spoke softly to me in a tone I'd recognized. The summer before, in Kansas City, Missouri, I'd visited Kellie and the family, and both he and I had gotten along great, minus a small discrepancy in our plans for his daughter. Recalling my last day in town, I had been sitting in the living room, sad to have to leave, when her father and mother brought us together to talk. There, in front of the family, they presented an engagement gift, saying, "We'd gotten this as a gift for your proposal, but since Barry hasn't asked yet, we thought we'd give it to you anyway." On that day, there was no confusion about what they wanted for us. Talking across the plastic receiver on the far end of a party line, he was making his expectations clear again. For several minutes, he went on counseling us about the sanctity of marriage and the need for abstinence until committing before man and God in unity. I did not disagree, but I did not have a lot to say. We'd already crossed that threshold, and it was time to make a decision on what to do next. Closing, he issued a final ultimatum.

"You both have an option," he said. "You need to either get married and end the sinful life that you are living or end the relationship that you have started."

After a few more minutes on the phone, we hung up, and I sat alone in the darkness of my room, searching for direction. In my search, I turned to God, but strangely, God could not be found. Truth be known, I had been confused about who God was for some time now, and my search to understand God seemed to get lost in the midst of all the religion that I was now surrounded by. Having taken a very literal path to God, I had embraced the notion that, word-for-word and verse-for-verse, the Bible had outlined my course of action. However, what I needed and wanted now was something practical, something that I could apply to this moment, but I could not find it within the dogma of much of what religion had to offer. While the structure and ritual made decision-making easy, it didn't tell me how to practically apply those decisions to this situation. It didn't show me if Kelly was the right woman for my life nor explain how to traverse a slower path and reverse some of the forward momentum we already had. It didn't offer the nuance of how to navigate my raging hormones or make a wise decision choosing the person I would spend my life with. It was merely a black-and-white guide of do or do not. Much like the ultimatum of her father, the rigid dogma of religion said, either get married or get out, and I was just trying to find a place to land the plane on a runway and think. What I needed most was the counsel of a loving mentor who understood exactly where I was and could help me through, but that mentor was not there, and I remained undecided.

The next day, Kellie called and ended the relationship, soon to move home. Having lost my first love, I was broken up. Determined to learn from my mistakes, I committed that if I ever fell in love like this again, I would not let indecision win over a decision. It was a commitment that I would keep and one that would send me on one of the wildest adventures of my life.

MY FIRST FILM, "A SIMPLE PLAN" (1998)

On the heels of love lost and being disillusioned with religion, I discovered something that I really liked. It was a stage but not a stage where I was the minister of a congregation. Instead, it was a stage where I could perform for the material interest of both myself and the audience. That stage was the stage of acting in plays, films, commercials, theater, and the like, and with it came a new direction and a new dream.

While there had been signs of this interest in my roles in three high school plays, it was never anything I considered viable until I befriended a woman named Liz, a producer in Minneapolis. Functioning as a pseudo agent, Liz had pushed me to get my first set of headshots, secured auditions in commercials, and helped me win my first job as a model doing an Aveda runway show.

Walking in front of a mob of people with cameras snapping shots, I finally felt like I was somebody of importance. Up here, all my failures seemed to melt away, and my peers looked up to me; it was significance without sacrifice, and I liked the feeling that came with it. But that dream would soon be stunted by a voice echoing deep inside me from a more practical side of where I'd come.

That inner voice, resounding like a parent to a child, reminded me that I was only a small-town boy who was a failure in school and relationships. The voice pointed out that I had a father who worked in industrial America and a mother who worked as a school bus driver. Inside, I felt ashamed of my desire to seek the spotlight and serve my self-interests. This was a career for those who came from money, and I came from little of that. So, the idea of going into the arts and film as a career was quickly shut down and pushed back behind the doors through which it had entered.

Instead of going hard toward what I loved, I started to live a double life. By day, I continued the charade of pretending I wanted to be a minister, while at night, I pursued my dreams of being an actor and working in film. During this period, I was split in two. I wanted to be on a stage to share love with people and entertain them but was too afraid to admit it to myself and voice it to those who had high hopes to see me preach. So, I continued my way through Bible college while working on commercials when I got the chance. Then, at the end of my last year in college, a moment changed my life's direction forever and inspired me to go after this new dream with a fervor that I had not previously known existed.

After having completed a series of commercials and theater in Minneapolis with the help of Liz, I became more confident, and bigger opportunities were coming my way. Then, I got a call that changed everything. She was working on a film in town that director Sam Raimi was shooting and wanted to cast me as an extra.

The name of the film was *A Simple Plan*. It starred Billy Bob Thornton and Bill Paxton. The premise revolved around the discovery of a crashed plane, a bag containing 4.4 million in cash, and the painful reality of greed on the most practical scale. It was discovered by two brothers, Jacob (played by Thornton) and Hank (played by Paxton) while they were walking in the woods. The brothers were local rednecks who, in their discovery, made a plan to hold the money until things settled down. But the plan was foiled when they killed a local farmer while returning to the scene of the crash. From that point forward, chaos ensued, and things quickly went south. In the end, Hank kills Jacob and turns himself in to the FBI. And no one walks away with the money. The scene I would be in was at the end, when the FBI finally discovered the crash and sent a crew of officers in to scout the area. I was cast as an officer and became excited about the big day.

In the middle of winter, I drove to Anoka, Minnesota, where cameras were set in place to roll and lights were prepped for

CHAPTER 3

action. I was in a wardrobe dressed from head to toe in an official county police officer uniform. In fact, I looked so authentic that on the quarter-mile walk to the set from the talent trailers, a family in a minivan pulled up next to me and rolled down the window. Driving through Bunker Hills Regional Park, they stopped and asked me whether the lake was thick enough for ice fishing. Conflicted, I knew I needed to break character and give myself up for their safety. Smiling, I informed them of what I was doing, and after laughing, they drove off to find an officer who could help.

As the winter evening shifted to night, the scene was set. Standing on a set in the middle of an airplane crash in snow-covered woods with camera and crew ready to roll seemed surreal, but with my hands and feet nearly frozen to the bone and the snow crunching to the step, I was reminded that there was something very real happening. After several angles and shots with myself in a mix of officers working the search area the director singled me out to be on camera. Walking toward him in my boots and uniform, I listened as he stood wrapped in a giant black puffy jacket, surrounded by a massive crew of people. Grabbing me, he instructed me to walk past the camera right and pan my flashlight up into the lens. Prior to that moment, I had heard stories about being "discovered" and had always dreamed that my day would come. Now, I began to wonder if this was the day.

"Can you do that?" he said. "Can you pan past camera?"

"Yes," I said, then returned to my first mark, my starting point in the scene, and waited for him to call action.

As they finalized lights, I couldn't help but dream of seeing myself on the silver screen. I imagined people soon knowing me by name. They would see my face and know that I was somebody of significance. In some way, I felt that this recognition would finally prove to the world and me that I was not the failure who had been cast aside in the third grade—they were all wrong about me, and I was now going to prove that I could be a success no matter what. While it was a completely irrational thought embedded in my deep-seated feelings of inadequacy, it was the driving force that pushed me forward in the cold Minnesota night.

On hearing the director call, "Action," I snapped into motion and performed exactly as he had asked. With an overwhelming sense of calm and focus, I stepped to the camera, looked past the big black glass eye of the lens, and walked off.

Moments later, I heard, "Cut, that's a wrap," and shooting was done.

Closing out the day and driving home, I wasn't sure about everything that had happened that night, but I knew that something was born within me. After all the years of wondering what I wanted to do with my life, I finally had discovered something that I loved, filmmaking. Now, all I needed was to figure out how to make a living doing it.

Standing with mom and dad for my graduation
from North Central Bible College (1998)

Wynonna Judd, Kmart Commercial (1999) -
Barry to the right of Wynonna.

"Be careful what you wish for, you just might get it."

—Proverb, Aesop's Fables

CHAPTER 4

TRUE HOLLYWOOD (1999)

BREAKING THE FOURTH WALL

Nine months had passed since my work on *A Simple Plan*. Since that time, I finished my degree, walked the line, received my diploma, and returned home to live in my parent's basement in Michigan again. Working as a substitute teacher during the day and a waiter at Red Lobster during nights, I was busy making money and figuring out my next steps post-college. In my search for a new direction, I called up Liz, my producer friend from Minneapolis, who had moved to LA. Listening to her raspy voice on my parents' new wireless home phone, she encouraged me to come visit, and I decided to take her up on the offer. Two weeks later, I tucked into a window seat, skipped across the West, and found myself looking out over the ocean in Santa Monica, dreaming of what was to come.

While I was admittedly an impulsive person, going to Hollywood had been a dream since the age of fourteen when I had a unique epiphany. As a young boy sitting cross-legged on my parents' living room floor looking up at our fourteen-inch RCA ColorTrak television, I was entranced in the movie *Rambo,* watching Sly Stallone as Vietnam vet and ex-Army Ranger fight his way to personal freedom. Pressing my hands into the cheap polyester-nylon woven carpet floor beneath me, I steadied my gaze as Stallone survived amazing stunts and battled his way through a platoon of local police and National Guardsmen. There alone halfway between a suspended state of reality and reality itself, it hit me: Rambo was only a character in a movie, and outside of the movie, the one playing Rambo, Sylvester Stallone, was the same actor who played *Rocky,* another childhood favorite. This meant that this guy was an actor in a film, a person doing a job, and if this was something a person could do for work, then anyone could do it, and anyone meant me.

Years later, old enough to leave the home, find a job, and travel two-thirds across the country for work, I was bumping along through clouds on the descent toward landing at LAX, and my mind could not comprehend the sheer number of houses extending across the valley visible in scale from my window. Coming from a town of only a few hundred people, I had never seen a city as populous as LA and could not fathom how such a place was even able to exist.

When people who are not from LA refer to LA, they think they are only speaking of the city of Los Angeles. However, the term "LA" also refers to the massive swath of land that is Los Angeles County, which covers 4,083 square miles encompassing eighty-eight incorporated cities, including the city of Los Angeles proper. The population of Los Angeles County is greater than that of forty individual states in the US; it is home to more than one-quarter of all of California's residents and is the most populous county in the United States.

Touching down at LAX and walking out of the plane to the sidewalk, I could smell the ocean air and see the palm trees all around. It was warm, the sun was out, and reality began to work itself into my senses. Calling my parents from a pay phone on a beach at the end of my first day, I shared my enthusiasm for the place. I could hear the fear in my mom's voice, but at this point

CHAPTER 4

in life, it no longer influenced me. I was on my path now and chasing dreams that were all my own.

Jumping off the phone, I needed to get some sleep. The next day was a big one—my first time working as a Production Assistant (PA) on my first project, an MTV music video with the rap artist Busta Rhymes.

Working freelance as a PA on sets in Hollywood does not take years of education. Nor does it take a ton of working knowledge of cameras or directing. It doesn't even take a full understanding of production, although there are a ton of moving parts on any given set and plenty of skilled workers. To be a PA, all you need is a willingness to do the job—no matter how dirty—and a contact who will help you get your start and call you for work when it comes in. At the moment, that contact was Liz. She was the one who had gotten me my first job, and I was off to the races.

As I walked onto the set for the first time, I was blown away. It felt just like what you see in the movies. Trucks pulled up and parked on open streets with people rushing in and out to grab gear. Cameras, equipment, lights, and props were all placed in strategic locations and appropriate angles to fulfill the vision of the director. Dollies rolled by on tracks preparing for shots, make-up artists danced through trailers, set designers pulled furniture and art from cube trucks, directors barked orders, and producers ran around with walkie-talkies tied to their ears, shouting, "Yes, on it!" It was mesmerizing and surreal, and I had no idea what part of all this was mine.

To stay busy and be useful, I spent the first morning as a PA checking in with every person on set, asking if they needed help. When they responded, "Yes," I would jump in. Whether it was running extension cords, grabbing props, or moving gear, I was down to do it. The fact was I had no idea what exactly my job was even supposed to be. I knew nothing of unions or roles on set; I just knew I was there to work. Meanwhile, the other PAs who actually knew their job descriptions went about their tasks of delivering paperwork and running supplies for the production crew, not even realizing that I was a member of their team.

THE UNKNOWN ADVENTURER

After the better part of a day just jumping in to help, I found my way to the production camper, where I got some clear instructions from the producer. By noon on the first day, I had figured enough out to know who to report to, what questions to ask, and what jobs were essential. With all the pieces in place, the camera was ready to roll, and everyone needed to be quiet on the set. Freezing in place, I didn't make a sound as I listened to the director called out, "Action!" With that, Busta Rhymes, who'd been standing just off camera, began walking into frame and rapping to his new song—smoking a joint, he blew plumes into a fisheye lens, and the light, ever so perfect, showed every little twist in the smoke curling up. It was mystical.

Captured like a deer in the headlights, I couldn't believe that this was work. For the next twelve hours, I continued doing more of everything that I could to support the crew. By the end of the first day, I was exhausted but hooked. I knew that from this day forward, I was going to do whatever it took to do this thrilling career.

After five days in LA, most of them on set, it was time to catch my return trip home, but first I needed to run by the office of HSI Productions in Venice Beach to get my check and say goodbye to Liz. HSI production did work with some of the most famous people in Hollywood and the world—from Will Smith to JLo (Jennifer Lopez) and Alanis Morrisette to the American rock band Red Hot Chili Peppers; this company had its finger on the pulse. Getting a start here was gold in this industry, and I saw the opportunity to meet people who might help me with work in the future, assuming I returned.

Walking in the doors early, I was greeting people in the office and finalizing paperwork when a Latino guy named Ishmael remembered me from the Busta Rhymes set the previous day. At the time, he was an associate producer with HSI, had seen my work ethic, and had taken a liking to me. He was outgoing and friendly, so we quickly hit it off and exchanged contact information before I left. As I headed out the door toward the airport home, he stopped me and said, "Call me when you get back in town, and I'll help you get some work."

CHAPTER 4

Then, as quickly as I'd come to LA, I left and was back home in Michigan, reliving it all like a dream.

HOME ALONE

Sitting alone in my parent's basement, I was writing in my journal—something I had started years ago. Now, I was thinking about what to do next. In my mind, I knew the clock was ticking. Those LA contacts and new relationships, while real, would cool off fast. Waiting several months till all the pieces were in perfect place was not a reality. Sitting there in the cold of winter, I made a list of pros and cons to help with the decision. I thought about how excited I'd be on a film set and the action-packed Hollywood scene. I thought about making a name for myself and all the accolades that would follow. This all excited me—not to mention I was passionate about it and, by default, good. The process of being around celebrities and managing the hype of the industry came naturally to me, which I couldn't say about much else in life before this moment.

On the contrary, home provided something I loved: family and friends. It provided stability and the comforts of small-town life. These relationships had context and history, and while Hollywood was exciting, it was also very fake. I knew going there would be a drastic shift in direction from the values with which I had grown up in the Midwest.

Torn between the two, I knew that I needed to make a decision and go forward. I had been caught in a space of indecision in the past, and it had not gone well. The reality now was that I was young, and as hard as it was to leave, I could not put off my dreams. In my ears, I could hear John Keating (played by the late Robin Williams) in the film *Dead Poets Society* ring out with the words "Carpe diem—Seize the day, boys. Make your lives extraordinary!" I wanted my life to be extraordinary.

Reviewing my list of pros and cons one last time, I thought for a second and then made my decision—I was going back to Cali.

THE RETURN TO LA LA LAND

Breaking the news to my parents was not easy. They didn't want to stop me but didn't want me to go. Unable to communicate their emotions, they were silent about their disappointment and support. On the day my car was finally packed, they had left on a weekend vacation. I just figure it was too hard to see me go. With the house as quiet as a bird, I flew the coop, pulled out of the driveway, and started west in my 1997 Grand AM and the $2,000 I had to my name—money that I had saved from teaching and waiting tables.

The drive west to California would be my first time crossing this section of the United States. While I had done some traveling (i.e., Minnesota for college, Florida for spring break, Colorado for my senior year), I had never traversed by car beyond the Rocky Mountains. So, I had no clue what to expect.

I planned to first head north to Minnesota to see friends from college. Then, from there, I'd head west through the Black Hills of South Dakota on the edge of Montana and down to Wyoming. Afterward, I'd travel across the tip of Idaho into Utah, where I'd visit a friend in Deer Valley before turning southwest into the Mojave Desert, through Las Vegas, and into the promised land of California. It would be an adventure within an adventure.

Cruising along, I quickly got through the Midwest, and soon the evergreens and maple leaves turned into the high plains. Before I knew it, I was winding through the Black Hills of South Dakota, and it was there that I first captured the deep sense of patriotism this region had for our country. Watching the flat prairies turned into plains with small canyons leading into the foothills, I was entranced while driving. The coloration of the land varied greatly, uncovered by millions of years of a watershed that reflected the yellowish-golden hue of the rocks and soil. As the sun set on my first day on the road, the land seemed to glow against the plains and plateaus, and I decided this would be a good place to pull over and rest.

Driving into Mount Rushmore, I visited iconic heads of presidents past and enjoyed this distant connection to our capital

CHAPTER 4

from halfway across the country. The grandiose sculptures required a massive vision for the future, and it seemed to parallel the vision of connecting our land from coast to coast into one constitutional republic. Farther down, I paused to see an equally grand sculpture of Crazy Horse, a Lakota Native American who fought to preserve his people's way of life but was killed in the process. Looking on with the sun on my back, a cool breeze hit as the moment gave a mix of feelings between the greatness of our Founders' achievements and the realities of lives destroyed to attain that.

Soon clocking nearly 1,500 miles from home, I began to drift into Montana "Big Sky Country." Its beauty cannot be discounted, and the wide-open spaces captured something within me. As breathtaking as it was, I was only there for a moment before turning southwest toward Wyoming. Transitioning from prairie to high desert, the land is covered with green sage as the mountains slowly walk up into the sky. Riding in my little bubble looking out at the world, I watched it all pass by under valleys that could swallow up the sea and over mountains that would reach the heavens. Mixed between it all were the random broken-down ranch homes, horses, creeks and white caps. It all felt like a place I've always been and could never stay.

Before I knew it, two days of driving had flown by, and I was a few hours outside of Utah. Now, I began to think about my destination, LA. While I had some plans to couch surf once I arrived, I did not officially have a place to stay. Reaching Salt Lake City, I decided to ring up Liz and start to brainstorm. At the time, Liz had been dating a guy named Chris, who lived in the San Fernando Valley during the week. On the weekends, he had been renting a sailboat in the Marina del Rey. Since they started dating, he began sleeping at her place and wasn't using the boat much anymore. On the phone, she asked, "Would you want to live on a sailboat?"

This was brilliant, I thought. A sailboat was a level of romance I had not thought about to this point. Responding, I said, "Yes!" Needing Liz to check with Chris, I knew I would hear back in a couple of days, but the deal was all but done.

I now only needed a place to crash till the boat was available. Ishmael, the guy from HSI Productions, had mentioned that if I was in a pinch when I returned and needed a spot to stay, he'd be happy to help. I called him, and he agreed to let me crash on the couch. With just twelve hours of driving left, I was set to go and on my way to LA.

VENICE BEACH AND THE BOAT

Pulling into town after the drive through the deserts and over the Sierra Nevada mountains from Las Vegas, I was filled with excitement. Los Angeles was overwhelming in scale, and I could feel the energy of the place all around. Driving down Sunset Boulevard, I turned south on the Pacific Coast Highway (PCH) past Santa Monica toward Venice Beach. Ishmael's apartment was in the heart of Venice Beach, and after navigating LA traffic for a couple of hours, I finally pulled into the area.

Venice Beach proper was known for being a little bit grungy but not overwhelmed with homelessness like in the new millennium. In 1999, it was full of artists, skateboard culture, and street entertainers. The beach was also a major tourist attraction, bringing in an average of over 200,000 beachcombing visitors daily. As I turned the corner near Ishmael's place and parked, I quickly began discovering why.

Just a block away on the beach, tourists wandered along the boardwalk. And the landscape was dotted with fantastical-looking characters in outrageous-looking clothes. Amazing basketball players were waging three-on-three battles, throwing down outrageous dunks on the paved courts. In the distance, you could find a fenced-off section where very big, very oiled, and very tanned men and women were pumping weights. This was "Muscle Beach," the home of guys like Arnold Schwarzenegger, Lou Ferrigno, and Jean Claude Van Damme. Beyond this were the sunbathers and surfers. And just in front of me at Ishmael's apartment was a three-story building with a mural of the American singer, songwriter, and poet Jim Van Morrison

CHAPTER 4

standing some thirty-feet-high, shirtless, and leaning to one side full of plentiful rage. It felt like everywhere I turned, there was something to see.

After settling into Ishmael's apartment a block from the beach, I peeked outside the window and saw Chris Tamburello, a character from the hot nineties show *MTV's Real World,* walking past. I could not believe where I was staying. It seemed fictional, and again, the small-town part of me was blown away. Sitting down on the couch, Ishmael offered me a joint of marijuana to help me chill out. Seconds later, I was stoned and sinking deep into his couch. It had been a long day with a lot of driving, and it was time to crash.

The next day, I called Liz. She confirmed that I could move onto the boat and asked if I wanted to go over to the Marina del Rey and check it out. The Marina del Rey is the largest man-made small-craft harbor in the world. It was once a wetland estuary full of sea life and birds that was dredged out to become what it is today—a home and glorified trailer park to over 5,000 boats and vessels from around the world. It goes all the way from Venice Beach to Playa del Rey near the LAX airport.

The aerial view of the marina from space seems like the outline of a tree with branches. Each branch made small peninsulas. Those peninsulas are called basins and are respectable slices of land that hold condos, hotels, and parking garages. On the sides of each basin are a hundred or more docks reaching out into the water, and each dock holds around twenty boats. The parking spots are called slips, and my boat could be found within that network of basins, docks, and slips.

Liz picked me up at Ishmael's in Venice Beach, and we drove over to the Marina together to sign the paperwork and scope things out. Upon entering Panay Way basin, I knew this would be my home. Around me, palm trees lined the median, and in the water were rows and rows of sailboats and power boats gently swaying in the tide. The ocean air filled my nostrils, along with a general calm of being near the water, and I had found my sweet spot.

Walking down dock 300, where my boat was parked, I could see from a distance the little Catalina twenty-seven-foot sailboat parked in slip zero-two-seven with its mainsail swaying left to right. Stepping onboard, I pushed back the hood to the gangway and walked down past the cockpit into the cabin. Looking around, I could see that it was a small but cozy boat and roomy enough for me to stand up by the entrance. Walking to the back, I had to duck down to miss hitting the ceiling before squeezing past the toilet to check out the aft V-berth sleeping area. Stopping, I took a seat to take in the little space and get a feel for it. It felt like a little home, like being in a camper at a park, and I loved it. I was sold on the idea and headed out. I couldn't wait to sign the lease and move in.

After a full day, I returned to Ishmael's place. He had just wrapped up the day. The two of us again sat on the couch and smoked a joint, had some wine, and started to chill. Sitting there talking, I couldn't help but notice that there was something about this evening that was different from the last. Ishmael was no longer sitting at a distant chair, keeping ample space. Instead, he was now sitting next to me on the couch and seemed to be looking for ways to move closer to me. At first, I wasn't sure if he was uncomfortable and just wanted to find a better seat. Then, it started to register; Ishmael was coming on to me.

I am not gay, and for me, this was probably much like what a woman feels like when she isn't attracted to a man who is attracted to her. I needed to figure out a way to avoid this and be polite. After all, I was his house guest. Not skilled in the practice of being direct, I told him that I was very tired. I stretched, yawned, and made it very clear I was ready to go to sleep. Finally, he got the hint. Closing my eyes and covering up, I knew it was time to get off the couch and get on to the boat.

FROM THE SAILBOAT TO THE SURFBOARD

Having only been on board the boat for a handful of days, I was still getting situated when I ran into a fellow boater named Eddie.

CHAPTER 4

Eddie was a thin, wiry, high-energy kid from Minnesota who had also come here to LA to act. He was a very open person, always accepting new friends and, on the same day we met, asked me if I wanted to go surfing. With little thought and lots of time on my hands, I said yes. Before I knew it, he threw two boards on his car top, and we were cruising down the Pacific Coast Highway (PCH) toward a spot north of Malibu called County Line.

Rolling into a dirt parking lot an hour later, on top of the car, we'd strapped down a short board for Eddie and a giant long board for me. Past the guardrail that separated the parking lot from the eroding twenty-foot drop-off to the beach, I could see blue Pacific waves rolling in from the horizon. Out in the water, there were a few raisin-like figures or surfers in wetsuits on boards floating in the distance. I'd never actually seen surfers in the water before other than on television and was amazed to see one firsthand stand and ride a wave.

Eddie unstrapped the boards and placed the massive ten-foot-long one—that I would be riding—on the sandy beach. After giving me a few pointers on paddling out and duck-diving, he said, "I will see you out there." Then, he was gone with his board, disappearing into the white water and out toward a lineup of surfers waiting to take turns catching the waves.

Picking up the behemoth of a board, I walked down to the water to see what I could make of it. Wading up to my waist, I jumped on the big plank and attempted to paddle. Instantly, my balance was off; I was all over the place, and nothing about this seemed feasible. Waves were pumping in ahead of me, pushing me back. From the viewpoint of lying on my belly, they seemed like mountains. Adding to that, I did not have a wet suit, and the freezing sixty-degree Pacific white water kept smashing over me as I was getting a head freeze and chilled to the bone. Still, something inside me was drawn to succeeding, so I kept paddling forward into deeper water.

Over and over again, I got hit by whitewash from the breaking waves, and over and over, I was tossed off the board. Slowly, after losing ground, I worked to get myself balanced and resume paddling. Just as I was about to build some forward momentum, I would again get hit by white water and have to start over.

I could feel fatigue in my arms and exhaustion in my lungs. It was hard to catch a breath when the water kept washing over you, and the whole thing began to look futile. Then, just as I was about to give up and turn back, there was a lull in the swell—a break between waves, and suddenly, I got enough momentum to make it outside the breakers up to the line where Eddie was waiting.

Outside the whitewater, things calmed down. The power and mass of the waves that were once beating me up now passed under me on their path toward the shore. I was now unscathed and able to catch my breath. Finding my balance on what felt like a Lincoln log between my legs took time. The board tipped in all directions, rocking left to right, forward and back. As it pitched and polled, with the waves, I was constantly getting tossed over one side or another at the slightest motion.

In the midst of the lineup, I was unaware of how silly I looked to the locals who surfed daily. Clueless, I kept at it, and I started to gain my balance. After several minutes and many flops overboard, I found a spot on the board where I could sit still and look around.

Picking my head up, I took the first look at the shore from the water. Around me were the cascading cliffs of the Santa Monica Mountains along the PCH. Turning my head to see the full panoramic view, I was in awe of the steep wall of land pushing up from the sea and ocean below. Looking down, I could also see kelp deep beneath me in the water as it swayed in wave-like motions back and forth with the tide. There, amidst the rocks and fish, was a breathtaking world about which I knew nothing.

Turning, I saw Eddie just down the lineup. He was paddling in and catching wave after wave before ripping out the back side and launching himself into the air. The excitement of his experience was contagious, and I now wanted that for myself.

Looking into the horizon, I saw what looked like a wave coming toward me—one that I could ride. Out of the distance, I heard Eddie yell, "Paddle." So, I quickly turned my board toward the shore and dug deep. Pushing hard against the water, I began

CHAPTER 4

to gather speed. Over my shoulder, I could see the wave coming closer and taking shape. Under me, I could feel my momentum pick up. Then, my heart skipped a beat as the bump that was the shape of the wave began to lift me up and drive me forward.

Suddenly, as if the hand of God himself reached down and started pushing, my speed picked up, and the board started to skip faster and faster across the surface of the water. Now I was no longer paddling; I was completely moving under the power of the wave. Instantly, adrenaline began to race through my veins. Lying flat on my belly, I could feel the spray of the water hitting me in the face. I watched as fish passed beneath the reef, unphased by my presence.

In some way, the push of the wave felt like the hand of God, divine. My board had now harnessed the energy of nature. That transfer of energy was infinite and would continue forward long beyond this moment. For a second, however, I was in the flow of that energy in a space of nature that was both separate from and part of everything around me. At the moment, it was beyond me how to fully comprehend this experience; however, what I did understand was that for a second, my entire world made sense, and I could feel the existence of something greater than myself.

As quickly as it had begun, the ride came to an end at the shallows of the beach. Back at the shore, my heart began to calm, and my mind began to settle. As the adrenaline stopped and I caught my breath, I paused to ask myself, "What the heck did I just experience?"

The question was full of the unknown space between science, energy, and the divine. It was a question I knew I would not fully answer; it was bigger than I could comprehend. But it was a question that I needed to seek some of the answers to, and the only way I knew to resolve those answers was to paddle back out and do it again.

MEANWHILE, BACK IN HOLLYWOOD

In those early years in California, surfing became a focus, but it wasn't yet the primary part. I still dreamed of fame and fortune and was chasing the idea that I could somehow find a door into the industry by first working behind the camera. With my days full of PA work, I continued to bide my time, knowing that something would happen soon enough.

CLIMBING THE RUNGS OF THE LADDER TO THE TOP

For the first two years, I traveled as far as South America for one of the most insane cocaine-ragging, hot girl-chasing commercial shoots in my life, a story that deserves its own chapter in the book. Along with that, I had filmed on commercials and music videos with celebrities Will Smith, Jennifer Lopez, Brooke Shields, Jewel, Jennifer Love Hewitt, and more. And I had started to make good money. However, there was one problem in all my success, and that was that all my work was coming from one person, which left me vulnerable to losing it all. Liz, my friend from Minneapolis, who had helped me start, was MIA at this point; she had ditched me because of a bad breakup, and that left me with one person to source all my work, Ishmael.

The reason that was a problem was that for the past two years since sleeping on the couch in Ishmael's apartment, his advances had continued. They had turned into a sort of sexual harassment and predatorial behavior that I didn't know how to avoid. No matter what shoot we were on, Ishmael found a way to humiliate me by slapping me on the ass in public or calling me into his office for a fake meeting and then trying advances like foot messages or back rubs. The whole thing had gotten out of control. In part, it was my fault for fear that calling out his behavior would lose me work, and instead, I was hoping that by ignoring it, it would go away. But every time I escaped one advance, a stronger one would come. I was afraid to assertively face it, which created a conflict between work and self-dignity. Then, on a four-day

CHAPTER 4

trip to Nashville for a Kmart commercial, the situation reached a climax that would change my life forever.

Waking up the day of the shoot, I was excited to get to set. Now, for the first time in my career, I felt like I was in a sweet spot and pushed myself to stand out. This wasn't acting. It just felt good to be recognized for doing hard work, and I was working with the best in the industry. On any given day, when someone needed a coffee, I would grab it. If trash needed cleaning, I'd pick it up. If they needed a fresh battery for their walkie-talkie, I already had one in my back pocket. It didn't matter what needed to be done—I did it.

To add to my growth, this was the first time I was being recognized for the success I'd earned. I was actually winning in this chaotically shallow industry and was starting to be known by name. The director knew me, the producer knew me, and on this shoot, even the talent knew me. The talent on this shoot was the Judd Family—in particular, Wynonna and Naomi Judd. During the day, they had noticed me working my tail off. As a reward at the end of the second day of shooting, Wynonna Judd invited me to join an after-party with the crew. I was flabbergasted and committed to finish my work early so I could make it.

Wrapping up my tasks, I headed downtown to the famed Whiskey Row in Nashville. This strip is full of some of the best places to go out. And on this night, Wynona had reserved a private space. At the front door, I was instructed to give the password to the bouncers. It was "turn it loose," the title of one of Wynona's best hits that she sang with her mom. These three words would get me in the backdoor, and no other words would work.

On my way to catch up to the crew, my imagination was running wild. This was my chance to start to fulfill my vision. I had always planned to find my way into Hollywood through the backdoor, and this was my first step. The director, Mathew Rolston, who casted work for some of the biggest commercials in Hollywood, would be at this party. Both were big players in LA and New York City. I knew that by hanging out, I would be one step closer to making the connection that I needed to get in front of the camera.

Walking up confidently to the entrance of the bar, I encountered two bouncers who stopped me in my tracks quickly. They were big, brawny, intimidating characters who would make anyone reconsider their decision to go inside. Standing there, I got close enough to share the password without having to say it too loud. Leaning forward, I dropped the words, "turn it loose," and instantly, I was golden. While crowds stood and watched, waiting to get in, the seas were parted, and I was rushed through to the middle of the bar toward the balcony upstairs, where Wynonna was sitting.

On my way past the bar, I could hear people whispering about Wynonna being up there. Some were even pointing and turning to see what was happening as I walked past and up the stairsteps to fame.

When I got there, Wynonna was pleasantly surprised to see me. She smiled and said, "I didn't know you were gonna be here, honey-bunny."

Her words were like a personal song written to me and me alone. I smiled back as the crew swept people aside to find a seat for me.

Sitting down, we instantly struck up a fun conversation. Halfway through it, she paused, and as lucid, free-flowing, and light-hearted as a southern girl can be, she turned and said, "Oh honey, if only I were twenty-three again."

Keeping up with the pace of the conversation, I replied, "Why do you need to be twenty-three?"—to which we both laughed.

Firing back, she retorted, "Well, I did just get divorced, and I am feeling a little loosey-goosey."

Pausing, I clammed up for a moment, and it was apparent that I was no longer sure if we were joking or serious. Without missing a beat, she noticed my concern and leaned back, patted me on the knee, and said with a smile, "Don't worry, honey. Mommy would kill me in the morning." Bursting out with more laughter, we both chuckled and then went on with an amazing night out.

By the last day of the shoot, I was burnt out. I had been pushing myself nonstop for nearly a week. On average, I got

CHAPTER 4

about four hours of sleep per night and was now running on fumes but was still loving the experience.

On the last day of the shoot, Ishmael invited me to join a smaller party in the hotel room of director Matthew Rolston and a few others. I was tired but didn't want to miss this.

As mentioned, I had planned to find the back door to get on stage. I knew that the majority of people in the industry were spending all their time working through agents and going to auditions. I felt like this was a massive waste of time. For me, the ideal way to get on stage was to get to know the people who were hiring, and Matthew was one of those people.

Matthew Rolston started his career as a photographer shooting beautiful models and celebrities for *GQ*, *Esquire*, and other fashion magazines out of New York. That career and his talent bled over into directing high-end beauty commercials in Hollywood. With his combined ability to direct and take photos, he quickly rose to the top of the industry. In a few short years, his work put him in the company of celebrities I've mentioned, plus others like Penelope Cruz, Minnie Driver, Will Smith, and more. By the time I started working for Matthew, he was making around $15k–$20k per day, directing commercials for the biggest brands and most beautiful women in America. I felt lucky to be one of his production assistants and, frankly, one to whom he had taken a personal liking. In my mind, this party with a very small, select group of people was a surefire way to get a little closer to my real dream of being in front of the camera.

After finishing up the work day in advance of the party, I needed to commute thirty minutes from my hotel at the Holiday Inn to the director's red carpet hotel at the Four Seasons. Quickly, I cleaned up, jumped in a cab, and rode across town. Then, I went up to Ishmael's room, where he'd told me to meet before.

Upon my arrival there, Ishmael, whose room had an open bar, mixed a few drinks for us. He served up a gin and tonic and handed me a big joint, then told me to hang out for a moment while he went to check on things. Agreeing, I sat down, lit up, and drank up as I waited for his return. Soon, my eyes were heavy, and I started to feel exhausted as I had been working hard, so it felt normal. Laying my head back on the pillow of the bed, I was out cold within seconds.

I am not sure how long I was asleep from that point or when Ishmael had returned to the room. I didn't hear him, but I do remember that as I was coming out of a deep drowsiness, Ishmael seemed to have moved me, and I was trying to sort out why he was straddling my back and trying to give me a back rub. My clothes were on, but he was slowly pulling out the bottom of my T-shirt and working on pulling down my blue jeans. Struggling to wake up, I tried to get him off me but couldn't completely come out of my slumber. Not wanting these advances, I started to swipe him away, mumbling several versions of "stop!" Then, it all became blurry.

When morning finally came, I was curled in a fetal position above the covers on the side of the bed with all my clothes on from the night before. Getting up fast, I wanted to get out of there. Whatever happened that night was unclear, but what was clear was this had gone far enough, and no job was worth this predatorial harassment.

Grabbing my stuff, I rushed out of the room, and within a few hours, I was on the road driving the cube truck back to LA. Caught in a deep sense of shame and embarrassment, it became clear just how stupid my plan to become famous had been. In the almost two years that I'd been chasing this dream, I had compromised my self-respect and given up my dignity in hopes of winning over the hand of the devil. Inside, I felt the empty shallow void that could only come from chasing the material world and being used like a cheap coin. Outside, I had no idea where I was going next, but I did know I was getting out of Hollywood.

Barry in the Marina del Ray next to his 27' Catalina.

The Jeep parked at a beach near Rio Nextpa in the ride from California to Costa Rica (2003)

"Better to be honest and face your fears than to lie and betray your dreams when fear shows up."

CHAPTER 5

THE ROAD LESS TRAVELED (2002 – 2004)

THE INNERMOST CAVE

Back home in LA, I was deep in the belly of my twenty-seven-foot Catalina, alone. In front of me was a mix of bright green crystalized marijuana leaves, stems, and seeds on a cutting board. I was working through them slowly and breaking it down into the most valuable, essential, and smokable parts. An hour or so earlier, I had left the comforts of my sailboat and visited the slip of an "old school" pot dealer who had John Lennon round glasses with pink lenses, a long gray beard with a ponytail, and a Hawaiian shirt and living on a thirty-six-foot powerboat; it was like stepping into the 1960s.

Stepping on board, I entered through his sliding glass doors and proceeded through the kitchen down into the main cabin. There, reclining and watching *The Simpsons* on a La-Z-Boy was a sixty-year-old man. Greeting me, he would always engage in conversation before pulling out his elementary tin square Star Trek-themed lunch pail full of all the goods. Inside was a variety of reefer that ranged in colors from purple and gold to plain green. It was like a scene right out of *Alice in Wonderland*; he was his own version of the Cheshire Cat.

This was before the age of lab-grown weed and the dark green stuff with stems and seeds was the traditional Mexican ragweed that fit my budget most days. Pulling out a scale, he would weigh out a kilo, stick it in a bag, and take my cash. Then, he'd pass me a toke on a new joint he'd just rolled, and I'd hang out a little longer before going our separate ways.

Back on my boat, I was working under candlelight and finishing rolling up a new joint for myself. Pulling out my lighter, I lit it up, took a nice big puff, and relaxed. Leaning back, I could feel the world slow down around me. One at a time, my thoughts collected as my emotions began to pipe down a line deep inside me.

Drifting there into my imagination, I began to draw up strong visual memories of my life since childhood. In this space, I began to have an out-of-body-like experience, and for the first time in some time, I saw myself from the perspective of an outsider. Like the days of sitting high on the branches of a tree and looking down, I observed my life and had a revelation that would end up forcing me to deeply review the core of who I was inside.

At first, I felt like I was diving down into a pool, swimming toward the bottom. Descending below the surface, I watched as my short time of living in LA quickly passed me by. Then, as I went deeper, I began to pass through my memories of college where I could see my indecision and look directly at my conflict with religion and my search for God. Beyond that, I plunged further down into high school and then middle school. In the midst of this, I saw a fifteen-year-old boy who had survived a near-death car crash. In this space, there was a great deal of fear and trauma—the pleading to slow down and the complete loss of control. Vulnerable and scared, I had reached out to God and could see myself crossing over the thin line between life and death.

Deep in thought, I was feeling overwhelmed and metaphorically desperate for breath, but I wanted to see more, so I pushed on. Below me, the water was dark and cold, but in the distance,

I could see a faint golden light glowing out of a cave. With only a few seconds left of air, I decided to push on, hoping to look inside it.

Paddling hard, I reached the edge of the cave and stood at the entrance. Peering within, I could see a small room, and inside was the boy sitting alone at a desk, working to keep up. It was me in my elementary classroom, and this was the boy who'd been left behind. Inside, I could feel the pain of isolation and the deep sense of being separated from the group that made up my peers. Emotions of shame, embarrassment, and loneliness were pulsating in waves and stronger here than I'd ever known. This child wanted to be accepted, validated, and reinstated by the institutions that had rejected him. Down here, however, there were no institutions, no teachers, only a boy alone working to catch up unaware that time had left him behind long ago.

From the surface, the cave had been only a glowing speck in the distance, but down here, it was a fortress with thick walls of fear designed to protect the boy in me from ever feeling these emotions again. Overwhelmed with what I had seen and needing to come up for breath, I pushed off the bottom and kicked to the surface and, in seconds, broke through. Drenched in emotion, I was finally awake and back on my boat. Exhausted from thought, I turned over in bed and drifted off to sleep.

THE PSUEDO FAMILY

After relocating to LA and living on the boat alone, I found it hard to make friends and had become increasingly more lonesome. While I spent many days and nights solo floating alone, I did happen to make one friend, Chris Korolczuk. He was the previous tenant on the boat and the ex-boyfriend of Liz who had helped me with my start in LA. Although Liz had broken up with Chris and, as mentioned, was MIA, I remained friends with him, and soon we started to meet more people.

Hanging out around our common interests in surfing, skiing, mountain biking, and hiking, we started to embark on road trips down south to San Diego or head up to County Line Beach to catch some waves. Over time, these adventures evolved into longer trips to ski in Mammoth Lakes and even Lake Tahoe. These excursions began attracting new friends to attend. Eventually, a group of friends formed that became a pseudo-family, with a bond even stronger than my biological one.

After two very lonely years in LA, I had formed a new family of which there were five members, including myself. Each member was unique and distinct in his own way. And collectively, we would come to resemble a small team of *Mad Max*-like, adventure junkies who loved to travel places and face new adventures together.

Among my friends, as I mentioned earlier, was Chris, a New Yorker from Long Island, NY. His visage was a version of Sly Stallone—that is, muscle-bound, punchy in his talk, with a hint of Long Island, and physically not someone you'd want to contend with. In hindsight, his Rocky motif might have been a significant factor in our friendship as my fascination with Stallone continues to this day.

The second member and the one who would forever alter my life was Dan Wasserman. We discovered Dan living in an apartment in Venice Beach, right next door to Chris. At the time, Dan held the position of one of the youngest campaign managers in the country, working for Congresswoman Jane Harman of the United States Congress. Hailing from Washington, DC, he had a strange habit of leaving his second-story bedroom window open during the day while he was at work. Interestingly, even before I truly knew Dan as a friend, we began engaging in a game of launching empty water bottles, crumpled paper, and food wrappers into his open, screenless window. This behavior, a contest to see who could make the most accurate shots, was a form of fraternal hazing and our way of showing Dan that we liked him.

Next came Dennis Stein, another member from the East Coast who had moved to LA from New Jersey. It was Dan who introduced the group to Dennis—a uniquely soft-spoken and very easygoing guy. He was also a surfer, soccer player, and someone with a poker face who was never easy to read. Although my skill set included strong intuition and an unusual ability to see through people, I couldn't see through Dennis. At first, this made it challenging for me to connect with him as I never knew where he stood. In time, I learned to understand him and found that he was the most sensitive and forgiving member of the group.

The final member to move to LA and join the group was Ty Bookman, a high school friend of Dan's from Washington, DC. The two of them had attended an Ivy League high school called Sidwell Friends, the same school attended by Chelsea Clinton. Dan and Ty had developed a relationship on the school's wrestling team. It was their mutual love of the sport that had deeply bonded them and became a recurring theme on our future trips. In the middle of a hotel bedroom, while they were dressed in tighty-whities underwear, an entertaining wrestling match would often break out between them. It was a rare and most humorous display of male bonding and tribalism.

Over the years, there would be additions like Riad, the funny Tasmanian devil of the crew, and Ricardo, the pot-smoking out-of-the-box artist. Then, there was Zac, Ty's younger brother, who was an out-of-this-world genius, or Ben, the mythical member who was talked about in infamous stories of high school drug use and shenanigans. Apart from these random cameos on the periphery, the original group of five individuals came to define my LA years and much of who I would later become. They would help reshape my perspective of self, drive me to go to places where I never imagined I'd go, and inspire me to discover things I never imagined I would find.

THE ROAD TRIP THAT GAVE BIRTH TO THE DREAM

Over a five-year period in LA, I had spent nights and weekends racing with this group to Mexico or north to surfing unknown places, but Monday through Friday, there was work to do. Looking around, I could see that most of my friends had started building careers, and even some from back in college had married, but I was floundering. After my jaunt with the film industry,

I struggled to find myself in terms of a career, and I had no clear purpose for my life. The fact that I was unclear on where I was going or what I was doing was tumultuous for me. I again felt like I was falling behind my peers, and it was eating me up inside.

In the years since the incident in Nashville with Ishmael, I had dropped out of the Hollywood scene and was teaching special education. I worked at a school for emotionally disturbed children in LA. But I was still genuinely unfulfilled. Outwardly, I worked to put on a good appearance as a functioning adult. Inwardly, I felt like I had given up on a big part of myself that I'd tapped into working in film.

To avoid dealing with the feeling of shame and lack of achievement, I again found escape in nature, focusing all my energy on surfing, skiing, backcountry trips, and the like. I started to mark off on the calendar in my classroom the days till the next trip, surviving work till the next three-day, four-day, or longer school holiday. Much like my students in the classroom, when the bell rang, I was off and ready to hit the road.

Then, in late February 2002, while at home doing my evening routine, I got a call from Dan that would change everything about my day-to-day and take my desire to escape in nature to a different level. Calling on the other end, Dan was enthusiastically pitching the idea of taking a three-month road trip from California to Costa Rica by Jeep. Starting by traversing the Baja Peninsula of Mexico, he wanted to cross over to the mainland, at Cabo via ferry, and traverse down Mexico into Central America with the goal of ending somewhere near Tamarindo, Costa Rica. The focus would be to find surfing spots along the way while camping and sleeping in tents or on the beach.

Dan had even taken the time to map out each of the epic spots we sought to surf. Promoting the adventure, he said, "We would hit Scorpion Bay to Puerto Vallarta, then Escondido to Witches Rock; we were going to escape society, live the dream, and surf until our arms fell off, then drive down the road and do it again."

Listening on the phone, I was overcome with excitement. Giving it no further thought, I impulsively said, "Yes!" But shortly after hanging up the call, I started to think of the feasibility of the trip and have second thoughts. I wanted to go on this trip, but to do so, I would need both money that I didn't have and a solid chunk of time away from my classroom responsibilities. In reality, I had rent to pay and bills to cover. While the trip felt like something I couldn't miss, the execution felt overwhelming, and fear took over my thoughts. So, as impulsively as I had said yes, I called back Dan and told him no.

In mulling over Dan's proposal, I thought back over my life. I'd never taken on something like this, and I didn't have an outline of what it took to make adventurous dreams like this into reality. My biggest dream had been Hollywood, which hadn't gone well at all. I realized from that experience that any dream, when acted upon without a clear pathway or road to success, can go painfully wrong. In some ways, a dream needed an institution of thought and the rituals of a religion to train, shape, and protect it. Without that, it was vulnerable to attack to survive.

In my home as a kid, I learned early on that dreaming and talking of dreams was far more entertaining than the work of acting on them. On occasion, around the dinner table, my parents would talk about their dreams of something new that they wanted for themselves. Sometimes, it was a new home or a new job. Usually however, these dreams would get lost in analysis-by-paralysis and ransacked by a fear of failure. Phrases would fly around the home like, "If only we had more money" or "I'm just not sure it's the right thing for now," and then the topic would be dropped altogether.

On occasion, I would get to see a dream hit the runway and work to take flight. Dad would come home with a new job offer he wanted, or Mom would find a new house she loved. On those evenings around the kitchen table, there would be a buzz of conversation within the family. In an instant, our imaginations would take off with excitement, and our heads would be full of how we could make it happen. Then, a day or two later, amid the new idea, self-doubt would well up, and uncontrollable far-into-the-future events would soon overwhelm the idea. Predictions of the worst-case scenario would be made, and before you knew it, that lovely little butterfly of a dream would be lost in a flutter of indecision. Time and time again, throughout my life, I'd watch this happen, and each time, we'd grow hardened from disappointment and more emotionally calloused.

The most challenging part of this was not the experience. It was recognizing that I, too, was as human as my parents and that the experience they were having was not unique to them. The fear of failure winning over the future was one that lived within me as well.

Just days after talking with Dan and after that flashback, I was back home in LA, reflecting on his road trip proposal. Inside, my fear of how I would make this trip work was still overwhelming. While I had called Dan back and bailed on the idea, I knew that no matter what, I needed to find a way to go. While I had no real-world experience of making a dream like this come true and was jumping headlong into the unknown, I decided I could go on a shortened version of the full trip focusing on the mainland of Mexico to Guatemala City—locations where I could fly in and fly out at two distinct points. Dan, who'd been counting on me as his companion, was initially pissed off and forced to find two other members—Zac and Dennis to fulfill the full distance of the ride; thankfully, he let me back in, and I was back onboard for a shortened version of the trip. Now, I just needed to begin packing my bags to go.

MEXICO TO GUATEMALA BY JEEP – DIVINITY WITH A WAVE

With the focus of a surgeon and the determination of a general, Dan had organized all the details for the trip, rallied the crew, and was ready to go. A short, bushy-haired, wiry little guy, he was very different from me in that once he'd made up his mind, he was not going back. Pulling out of LA for the first leg of the trip, he would be joined by young Zac, a wide-eyed new college graduate who was ready for the road. In two weeks, I would meet him in Puerto Vallarta, and Dennis Stein, the silent mediator, would join us in Acapulco.

CHAPTER 5

For the first two weeks of the trip, Dan traversed the mysterious and baron Baja Peninsula through the border town of Ensenada and down to the tip at Cabo San Lucas. Winding through the desert roads, he drove across rocks and small boulders from beach town to beach town, past mesas, salt flats, and the crystal blue waters along the coast. Always a promoter and great at rubbing in when you'd missed out on something, he made sure, once we finally saw each other weeks later, to share that he'd gotten it good. Hitting up isolated surf spots and camping on the desert beaches along the coast, he paddled out at the spots K36, K58, Abreojos Ojoyos, and Quatro Casa and bragged about reaching the most challenging spot on the journey with never-ending waves at the renowned Scorpion Bay.

After finally making the flight south to Puerto Vallarta, I heard about all of Dan's adventures and was painfully jealous. On landing, I was standing curbside just looking up from my paper map of the route when I saw a Jeep rolling into the airport to pick me up. Inside the gray-worn Cherokee with surf racks and what looked like a pile of junk strapped on top was Dan, a sight to behold. Looking closer, I could see surfboards, fishing poles, spare parts, and even a spare tire suspended on the roof and secured in sections with a blue tarp and yellow nylon rope. It all seemed like it could fall off at any time.

After pulling up rapidly to the curb, jumping out of the Jeep, and hugging me, Dan looked at the size of my rucksack full of clothes and the surfboard and gave me a good tongue-lashing.

"I told you not to bring all that gear, bowski," he yelled.

Looking at the lack of space inside, I could now understand why. With backpacks crammed into corners, a box of books straddling the center of the backseat, empty cans of food stacked in boxes, a guitar in case, and random pieces of surf wax strewn all around the floor, I could see Dan's conundrum. There was no space left to sit.

With traffic buzzing all around and taxis moving in and out, we hurriedly strapped my board on top and jumped in the Jeep. With Zac sitting shotgun, I gave him some love, and then we rolled out of the airport and soon headed down the road to bigger things.

Over the next few weeks, we would traverse down to the likes of some of the biggest and best surfing of my life. Spots like Rio Nextpa, Playa Linda, and Puerto Escondido were all firing—mostly out of control for my skill level, but I still paddled out. In the past four years since I'd first learned to surf at County Line, I had found my way to the beach multiple times a week. Although my skill level had moved from novice to advanced, I needed to be an expert or a pro to step out into some of these waves.

More often than not, after making the paddle out, I posted up and watched the show. These were some of the best rides on earth—watching guys dropping in off the "lip" of a powerful section of a racing wave before cutting into the face of the ride was "heroing" to see. Here, guys were dropping in on giant hollow waves at the last second just before they would pile drive them into the ground. With no time to spare, they would make the drop and disappear on a rocket, all before getting shot out of the other end of a barrel with mist at their tail. Time and time again, they'd come firing down the lineup. And time and time again, I thought if they can do it, it can be done, and if it can be done, I can do it.

My competitive spirit craved to catch up. I wanted this more for myself than I'd ever wanted anything. Day in and day out, back in LA, I'd trained on waves to get to this level. Still, I was not quite there. Thankfully, throughout the trip, there were more manageable spots where I could train and hone my craft, ride my waves, and still enjoy the beauty of being stoked—a thrilling measure of excitement that cares not for the size or speed of a wave but only for your enjoyment.

After a fun day of riding and getting salty, I would traverse back to land and take in probably the most favorite part of the ride—eating tacos and hanging out with the locals. On every portion of the trip, from northern Mexico into the deep south, the people were genuinely the best part—always friendly, always cooking good food, and welcoming in their simple, natural ways.

CHAPTER 5

CONFLICT ON THE ROAD TO THE PERFECT WAVE

Away from the surf and off the road in dirt parking lots and camp spots, we'd had our share of drama. Putting four testosterone-filled, twenty-something-aged men in a small space with lots of heat and little sleep was bound to push things in many directions. Dan had not completely forgiven me for dropping out of the full trip after I had promised to initially join, and he showed it with some of his punishing remarks and isolating behavior. I'd sought refuge in my friendship with Zac and Dennis; however, outside of the waves, the group began to split with the tension reaching a head on a beach in Rio Nextpa.

There on the sand, Dan and I had broken into a full-blown argument, and I was ready to fight. Racing around the back of the Jeep from the passenger's door, I met Dan standing atop the driver's side door step, looking down. Dug deep into the sand, I was breathing heavily and prepared to charge with fists at the ready. There on the beach with the sun blazing overhead, a blunting rage of words fumed out of my mouth, the crux of which now revolved around what was to happen with the Jeep—a vehicle promised to me upon the trip's end that Dan announced was never coming out of Central America.

In an explosive moment, I verbally blasted Dan with my frustration. Internally, I was distraught that I'd let him and myself down by not fulfilling my original commitment. Externally, I was enraged that he wouldn't forgive me and let it go. Unable to manage my temper, we faced off on the sandy beach until cooler minds prevailed and Dennis interceded to calm things down. With tension in the air, we separated and let things begin to pass.

Adding to the tension was the general discomfort of traveling the country in this fashion. For days and weeks on end, our clothes went unwashed and were dirty and stinky. Our beards were long and itchy. The car and the tent stank horribly from our body odor. Nights on end, sleeping in tents or hammocks outdoors with bugs and critters waking us up was wearing us out. However, each of us had become accustomed to living with a certain amount of exhaustion from the incredible heat and humidity and overall lack of sleep. With my departure just around the corner in Guatemala City, I had five days left on the trip, and there were rumors of an unknown spot that was echoed to have epic surf. I was hoping that before I returned home, I'd get that perfect wave to make it all worthwhile.

Beyond the drama and struggle, we still needed to drive, so we headed deep into the state of Oaxaca, near the southern border of Mexico, searching for a magical spot called Chicahua.

The name of the place was just hearsay, and the location was not on any road map or guidebook. It was not in the publication of *Surfers Magazine* nor talked about in the circles of the World Surf League (WSL). This spot was a location that could only be found or discovered by working and exploring unknown places and spaces. This kind of search for waves felt divine to me and was the very essence of what surfing is all about. It is a search for something you've never seen and for a venture that you have never experienced.

Driving into the town of Los Tamarindo along Mexico's Highway 200, Dan began to ask locals in his broken Spanish if they had heard the name, Chicahua. One at a time, we were rejected and turned away. Each time, we'd drive in small jaunts farther south until, finally, in one small village, we got a sign. Watching from inside the Jeep, we could see Dan talking to an older man on the side of the road. Around us was nothing but dirt fields and jungle. In the absence of any road, a man turned and pointed off into the distance between two fields. Somewhat uncertain of the direction but confident enough to start the trek, Dan hopped back in the Jeep, and we turned right and started to drive down a dirt country lane in the direction of mountains and the jungle.

Rolling along over bumps and through mud holes, we had no map, no GPS, and no clue if we were even headed west. In the chaos and uncertainty, Zac began to call out the names of the capitals to states across the US, and we all joined in on the challenge. Sounding off, he would call out the state, "Arizona," to which we'd call out the capital, "Phoenix." This practice entertained us for some time as we worked through the fifty US

CHAPTER 5

states before moving on to countries. Soon, fields began to transition into the shallows of the jungle. Small farm homes with chickens and goats grazing on the roadside grass popped up, and we slowly transitioned into the full canopy of the jungle. Still, without a clue, we continued cautiously onward toward the darkness under the deep foliage and advancing toward some sort of an end of which we knew not.

After at least two hours of driving, we found ourselves facing massive water holes left over from a passing hurricane a few days prior. Not wanting to get stuck out here, I was forced to walk in front of the Jeep, testing the holes for deep mud or deep water. Doing so, I began to wade into the brown, dirty waters, testing their depth and slickness while Dan followed behind via Jeep.

Traversing deeper, we started to see the traces of a few "Halloween crabs." Showing up as only a few at first and then ultimately blanketing the road, these crazy-looking things began to emerge from the shadows covering the ground like Amanita muscaria mushrooms. White and red-looking things jutted and zig-zagged back and forth on our path until, at the last second, randomly scrambling out of the way, missing the bottom of the right and left tires by inches. Finally, after what seemed like a test of will and fortitude, the curtain of the jungle receded, and in front of us was the wildest of villages, full of homes built of cinderblock and mangrove twigs.

Peering through gaps in the open structures, you could see people near wood fire ovens watching who was passing through. Inside, families were sitting on the ground or in handmade chairs cooking in a brick oven. This was one of the most rudimentary communities I'd seen to this point on the trip.

Pulling into the center of the village and parking, we were soon surrounded by local boys and girls excited to see strangers with white skin. Hot, sweaty, and tired, I walked up to them with the friendliest of smiles. In doing so, I noted that the villagers had a darker complexion than most Mexicans of European descent and hair that was much like an afro. These were true natives of this land, very isolated from the influence of the outside world.

Surrounded by kids, we snapped a photo but soon needed to reach our final destination. By this point, evening was starting to settle in, and with it would come complete darkness. While connecting with the locals was fun, we still needed to cross a lagoon that separated us from the ocean and the surf that we had come in search of.

In our travels to these remote locations, there were no hotels to call or places to book online. All planning happened on the fly. However, while mixing and speaking Spanish with the locals, Dan had learned of a man named El General on the other side of the water. There, he owned two huts, and the chances were good that they were available for the night.

Asking around, we found a local who would provide parking for the Jeep at one peso per day. We agreed, and afterward, he escorted us to a ponga to ferry us over the lagoon. Quickly, we parked and were ushered onto the boat just before dusk. As the boat skipped over the glassy waters of the lagoon, we watched as the distant lights from huts along the shore began to appear.

With the faintest of moonlight, we could see the silhouette of the palms as we rode closer. From the boat, it felt like we were in a scene out of the film *Apocalypse Now*, going deeper and deeper into the jungle in search of Colonel Walter E. Kurtz. Pulling up to a dock, we jumped off onto dry land and were quickly ushered to the property of El General. Walking from a house in the distance, we saw that he was a thin, lean, and short man. He greeted us, politely welcomed us to his property, and then guided us to the huts with beds, and we were happy to be able to crash for the night.

The next morning, as the sun crept through the cracks of the wood-planked, windowless space, we rose to see what was around. Pushing back the creaking door, I peered outside. From here, I could hear the ocean waves breaking and see the breakfast palapa on the beach.

Marching up to the large general hut, an open-sided, thatch-roofed structure, we all sat to watch the waves. Inside the kitchen, El General was working behind the stove. There, over a fire, he was brewing some freshly picked lemon tea leaves that he'd

CHAPTER 5

grabbed on his way up from his house. Throwing them in the pot, he added some sugar and then poured it into four coffee cups for us to enjoy.

Waking up oceanside, we could see the lines of waves rolling in off the jetty wall that lined the lagoon. Just out past one hundred yards were perfect six to eight-foot "sets"—a series of larger waves. They were breaking perfectly outside and peeling for what seemed like a mile before closing out far from where they'd begun. Instantly, our adrenaline shot up, and we needed to get out before anyone else came.

Throwing back the tea, we quickly jumped in our surf trunks, threw on our rash guards (lightweight, stretchy athletic shirts), grabbed our boards, and paddled out.

The surf break over the sandy seabed at Chicahua is a long right (meaning it breaks right-to-left), with a strong, powerful "shoulder" and a few hollow-barreling sections. Formed by the sand bar at the mouth of the lagoon, the wave wraps inland over the break that was created from the flow of the lagoon's tide. The takeoff to the wave is just along the jetty wall's boulder edge, making it feel like you could get sucked up and crushed. If you make that drop and get onto the face, you will experience one of the best rides on the Mexican coast for several hundred yards before closing out on the beach at the other end of the village.

Racing, I quickly ran down the beach with Dan and Dennis in tow, then paddled like hell to get outside. After a few duck dives and a long swim later, I wound up in the lineup with only my friends in the water with me. The wind was dead, and the surface of the ocean was calm and glassy. In the distance, the swell was rolling in toward shore with twelve to fourteen-second periods between sets of larger waves. Slowly, the waves crept toward us until they hit the sandy bottom and grew exponentially in size. Soon, the shape formed into a smooth, fine A-frame-shaped wall with its thin transparent peak fluttering in the wind.

Now it was time to take off, and Dan had priority. Paddling, he dropped his foot back on the board to drop onto the wave with his face tight to the wall. Disappearing into the wave from behind, he was gone for the moment. Then, a split second later, you could see his silhouette flying down the line through the back of the thin wall of water before his head popped up over the lip. With a turn, he'd spray a peacock's tail of water before disappearing again. After what seemed like an eternity, you could see a small raisin-like object pop out of the wave far in the distance and let out a hoot.

Sitting there, I watched as the next wave was served up to Dennis, who was a goofy-foot, like me. He dropped in with his back to the wave and disappeared. Ripping down the line, you could feel him pumping all the way to the beach.

Then, after the two of them had caught their respective waves and were paddling the long distance back out to the lineup, it was my turn to go, as Zac had not made it out to the water. Alone waiting, I stared into the horizon, searching the water's edge for a bump and my chance to ride.

Thinking back, I recalled the hard work of learning to surf. Days of driving down the coast to find the right spot and, time after time, working to first stand up, then turn and ultimately ride deep into the face of a wave. From the first ride on, I have tried and failed an uncountable number of times. Occasionally, I would succeed, and that would lead to a good ride, and each of those rides stacked together over time would lead to learning something new. Finally, after years of hard work and overcoming failure, I was at a new personal pinnacle of the sport. I had been riding every day for a month and was really starting to shred waves.

Wiping the saltwater from my face, in the distance now I could see a bump on the horizon, a wave beginning to take shape in the distance. Turning to paddle, I could hear the hoots and shouts of Dan and Dennis yelling for me to go as they paddled in my direction. Pushing hard, I suddenly felt what I've referred to as the hand of God pick me up as the energy from the water elevated me and moved me from behind down the face of the wave. Instantly, the board started to surface as I pushed up to my feet and turned right down the face. With the wave at my back, I began to pump into my board for power and speed. Suddenly,

CHAPTER 5

I could feel the force move me as I began to rush down the line, ripping up and down the face.

I was like the composer of an orchestra directing the flow and energy running through my body as it returned it to the wave. From my head to my toes, I could feel the might of nature. And in that space, something divine and holy passed directly through me. In front of me was the face of God, next to me was all creation, and behind me was the impermanence of time collapsing into the destruction of the white water with every experience I'd ever known.

Ripping along, I raced to stay present and in the moment while still maintaining an intense focus on what was coming in front of me without letting the fear of what was behind to pull me down.

Then, in a second after an experience that seemed like a lifetime, the ride was over, and I kicked out of the wave, landing back into the water. Unable to help myself, I yelled out with excitement. Then, wanting more, I turned to paddle back out to the lineup and my friends.

Ironically, the three of us would not see or catch another wave that day. The tide shifted, the current of the lagoon changed, and with it, the swell and the waves died. It was as if we had been given a gift—just one special wave for each of us and all our hard work before going home.

Driving toward Guatemala City, I relived that wave and the trip over and over in my mind. That epic ride had encapsulated so much more than I could possibly express in words. It was the beginning, middle, and end of the trip. Taking off toward home, I'd made amends with Dan and watched as he and Dennis drove south onward to the rest of Central America. Zac, ever living in his own world, was to jump on a bus and travel a path of his own.

Never again would we repeat this same moment in time, but this trip and those waves had grabbed onto me with something that I would never fully escape. That something would drive me deep into the unknown, where I would lose everything I held as important. In that space, I discovered a purpose that would define my life for years to come and help me recover from a life that was lost in the mundane and repetitious, a place that few ever escape.

The take off at Chicahua with Dennis looking on.

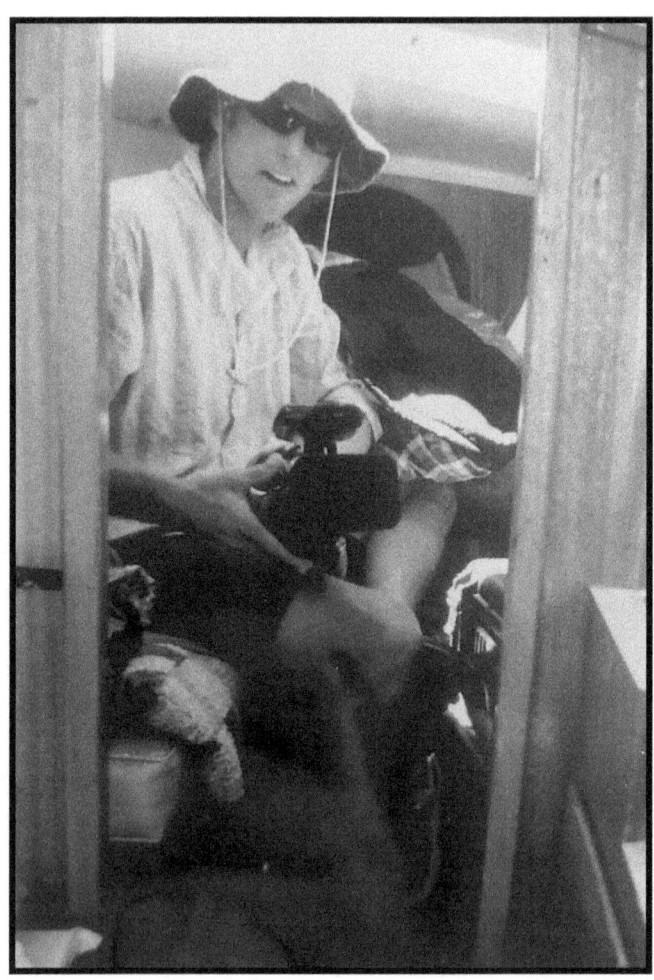

Barry onboard Rhino in the v-birth with his Soney FX1 Camera.

*"If you follow every dream,
you might get lost . . ."*

— Neil Young

CHAPTER 6

THE VOYAGE TO SELF-DISCOVERY (2004 - 2006)

SOCIETY, ROUTINE, RESTLESSNESS, AND WONDERING

When the trip down to Mexico came to an end, the group that had been my pseudo-family in LA for the better part of five years broke up. During the summer away, Chris fell in love with a girl from Long Island and moved back to New York, where he was from to settle down. He ended up being the first of the group to get married. Ty, also in a serious relationship, took a position in a company that had him traveling and staying in Europe. Between the girl and work, he was incognito more often than not. Dan, who had been the glue in many ways, moved to San Francisco to be with his longtime girlfriend, Adrienne. Dennis moved to Redondo Beach a solid hour away in good traffic from the San Fernando Valley and would soon be married too. And I returned alone to Reseda, California, and back to teaching.

With everyone gone and the adventures drawn to a screeching halt, I was feeling alone again. In a way, my phobia of being left behind was a self-fulfilling prophecy, and much like the child in elementary school, I felt lost. Floating around poolside in my Northridge home and gazing up into the blue Californian sky, I relived the experience of traveling through Mexico for two months with three of my closest friends. That trip had deeply impacted my life. I had completely unplugged from society and, in doing so, found something inside myself that made sense. The constant forward movement, the lack of attachment, and the ongoing change in the environment all made me feel alive. With that all gone, I felt dried up, looking for my next fix, like a soldier seeking his second tour of duty.

Without realizing it, I had become something of a travel addict, and nothing in my day-to-day existence could satiate it. Holding down a regular job felt like a mundane repetition that could not satisfy my desire to go on a mission. The world around me felt blasé, and I struggled to find purpose and meaning and could find nothing to solve this problem except to go back on the road and disappear into the wild.

A year after the trip, starting on my thirtieth birthday in 2004, I decided it was time to take action. Having spent the previous several months alone, I could no longer withstand the lack of adventure, and I started going on my own smaller adventures:

- In April, I summited Mt. Whitney (14,505 ft), the highest peak in the continental US, solo.

- In May, I drove out to Joshua Tree to the solitude of the desert to think for days.

- And every weekend, I had been driving up and down the coast in search of surf, my friends, and all that had once made sense to me.

However, I consistently returned home to my lonely self.

CHAPTER 6

Unable to calm my restless spirit at the end of the 2004 school year, I decided to sell everything I owned, resign from my job, move out of my apartment in Reseda, and just drive. Now homeless without clear direction, I set off on an undefined road trip with no specific goal or end.

From my home in Reseda, I first drove south the full length of Mexico's Baja Peninsula with a female companion, Leslie. Seeing firsthand the mesas, deserts, salt flats, and surf was surreal. All of Dan's recollections and rub-ins from his time down here were true, and I even found my way out to every spot he'd surfed, getting them equally as good. This checked the box, and in my mind, it made up for the first portion of the trip with Dan that I had missed, and I turned back north for new ventures.

Rolling along the tarmac, I headed north and drove Highway 1 along the entirety of the west coast of the United States through Oregon and Washington, crossing into Canada. There, I turned right and headed east on Canada's Crowsnest Highway 3 along the southern border through British Columbia and Alberta. An old mining and logging route, the highway wiggled across the back of beautiful mountain ranges and giant arch bridges. Driving over Kootenay River Gorge, my knuckles seized down on the steering wheel as I looked over the 500-foot drop of water below. Knocking out half of the west, I drove till I came to Glacier National Park, then turned south. There, I headed to Yosemite to see Old Faithful, then on to Zion National Park and into the Grand Canyon. From there, I again reversed direction and drove the backbone of the Rockies before turning east through the Dakotas, Minnesota, Wisconsin, and down into Michigan. Driving from top to bottom and left to right, I traveled until I reached the coast of Maine. It was a classic all-American tour from coast to coast for over 10,000 miles. I drove and drove and drove until the money ran out and the gas was all gone.

Broke, I returned to my parents' house, and like a wave that I had been running from, depression sat in—I was back in the very same place I'd left before moving to Hollywood for fame and fortune five years before. Needing to recover and get out of their place, I found a job back west. In January 2005, I returned to California to teach special education in the mountain town of San Lorenzo Valley above the small town of Santa Cruz. Moving into a cabin under the giant redwood trees—I was in for quite a change from anything I'd experienced before.

After having blown things up traveling for months on end, I was struggling with depression and working to make sense of my new life. While focused on putting the pieces back together, I couldn't help but want to reach for that euphoric non-stop motion of travel that had satisfied a consistent rush for me, a rush that the life of an average person was not fulfilling.

While teaching special education at San Lorenzo Valley High School in Santa Cruz, I was far from the world I had once known in LA, and I was again working to make sense of my life and figure out what I was doing with it. All the travel and disconnect from the world had driven me to the extremes of myself. The process of stripping everything away disconnected me from reality and forced me to face who I was inside. I was afraid of looking directly at who I wanted to be and going toward it, and everything that I had been doing to this point was to maintain the facade of what I wanted to be without actually taking the risk to get there.

In addition, the process of breaking away to learn these things about myself also left a gap between me and society. During the travel, I had broken away from my culture. I had separated from materialism, the nonstop consumption, and the endless race to earn more. Now, I no longer wanted to buy into it wholesale. The separation had given me perspective, and I did not want to lose that. So, I struggled to immerse myself back in it again. As a result, the transition back was hard going.

The return to work and routines coincided with my first Northern California winter in the mountains. In the cabin under the woods, there were no sporting events, no beers with the boys, and no one at home to keep me busy. While I still loved to go out for a surf in big seas, doing it alone had run its course. I needed someone to share it with, but I was now alone with only my thoughts. Towering over me were the giant redwoods. Far above them was the light of day blocked out by massive trunks

CHAPTER 6

and thick greens. This was a temperate rainforest with trees that drank a lot of water. Soon, the winter rains drawn from the storm waters of the Pacific began to pour down on the giant conifer trees and shrubs below with the essence of life. Pressing myself into the bay window in my cabin, I watched it come down and had nowhere to go. To walk outside in the rain under the deluge was like swimming. And in the damp darkness of winter, I began to fight my way back to the surface, back to routine to sober up from my travels and work to get back into the day-to-day of life.

Listening, I could hear the voice of my mother with her midwestern values, saying, "All this undefined time and travel was not what we were here to do; we were here to build a family and work."

I had been doing neither, and I could feel her disappointment even in her absence.

As the more traditional and conventional part of my psyche took over, I shut off creative ideas and dreams. The hopes of Hollywood were now far, far away. The ideas of filmmaking had passed. My whimsical, impulsive drive to jump off high places was being shut down for the practical and responsible. My history of poor decision-making had led me to repeatedly start over and over again—a form of self-sabotage. Now my left hand was holding back that impulsive right. Inside, the metaphorical Mr. Hyde was kicking and screaming while on the outside, Dr. Jekyll was doing everything he could to remain calm and build a life that Mom would respect.

As my first year of teaching turned into my second, I started to finally gain some stability. I met a girl in San Francisco I liked, moved out of the darkness of the redwoods into a sunny apartment by the beach, and was making new friends. As a result, I finally had a little money in the bank. However, while I was outwardly putting my life together, inwardly, there was still something missing.

While teaching paid the bills and satiated my love of the stage, at some level, it also felt like the easy road inside. I felt there was still more to prove to myself—the child in my inner cave desired validation. And the adult on the outside wanted a new mission.

Battling between stability and a lust for adventure, haunted by a part of me that felt suppressed and watching time tick by on my wall of life, I began seeking a resolution to my conflict and soon found it.

THE SAILBOAT, THE SEA, AND ME

In the middle of the week, I got a call from Dan, who was living just north of me in San Francisco. On the other end of the phone, he started to pitch a new travel adventure, one that would surpass all others. Instantly, my ears perked up, my wheels started to turn, and everything I was doing to build stability seemed secondary at best.

For the previous two years since the completion of the road trip through Mexico, Dan had been living in San Francisco. He, too, had been working to settle into a normal life but was struggling with it. He had an office job and was living in a small apartment with his lifelong girlfriend, Adrienne. Dan, similar to myself, had rebuilt his life into a traditional routine of commuting, socializing, and dating like so many others in the culture. But like myself, he had internal turmoil and angst about living a "traditional life." Inside, he wanted to do more than just pushing pencils and sitting in meetings. He wanted to be out experiencing life while he was still alive.

To help sate that desire, he started looking at sailboats. Some years earlier, on the drive through Mexico during a beer-soaked night, Dennis had mentioned the idea of a sailing-surfing adventure, and that had stuck with Dan. One day, while sitting on the toilet flipping through a sailing magazine, he saw a photo of a boat anchored on an island along Mexico's Baja Peninsula. In that picture, on a calm, sunny morning north of the anchorage near the boat was a peeling ocean wave with nobody on it. Sitting there on the throne and reminded of Dennis's comments, a dream was reborn, and Dan went into action.

Within a month of that moment, Dan purchased a twenty-four-foot Islander Bahama boat for $2,000 and was teaching

CHAPTER 6

himself to sail. Then, he started calling Dennis and me, the only two friends he knew were crazy enough to join him. Picking up the phone in my apartment in Santa Cruz, I listened as Dan promoted the idea. He was rambling emphatically about the shortness of life and the few opportunities we have to really live it.

I was a sucker for this type of grandiose rhetoric. So, the need to "seize the day" felt imminent within me. My emotions to go with it were irrational, unthought-out, and impulsive—and for the first time in my life, I could see myself getting sucked in and realized the impulse brewing.

While he rambled on, I drifted off, realizing that this was not the first time Dan had used this tactic on me. He had made a similar call to action for the road trip through Mexico. At that time, I was so overcome by excitement that I impulsively said yes before hanging up. In retrospect, that experience had been a source of contention between us that had boiled into a near fight on the beach—an undertone of energy that quietly lingered on to this day. Although I finally made the voyage to Mexico, my impulsive response and subsequent flip-flopping on commitment had caused me to miss out on Baja and Costa Rica—making it feel incomplete. Those were precious moments I could never get back. And I'd promised myself that I would not make the same mistake again.

Listening to Dan finish his pitch, I told him I needed some time to think. Then, I hung up. Alone in my apartment in Santa Cruz, I began to mull over my life, routine, and commitments. I started to think about all that I would need to do to make this work. In the past few years, I'd been on multiple month-long trips. Every time I returned from my travels, I had to go through a painful transitional rebuilding period that took months and even years to adjust to.

Since my previous travels around the US and down the Baja, over a year had passed. I finally had a stable job, an apartment, a girlfriend, and some money for myself. As much as I struggled with the mundane routines of life, they helped stabilize my emotions and pull me out of a deep depression. However, life felt like it was supposed to be about more than just routines. It felt like it was supposed to be about experiences and dreams. As a matter of fact, despite all the routines, there was one dream that had been dropped but still lingered inside.

Over the years since leaving Hollywood and the film industry as a whole, I felt incomplete. While I found solace and income as a teacher, I had not found the passion and purpose that I had experienced on the sets of films where, surrounded by famous actors and directors, I felt significance. For the first time in a long time, I knew what I wanted to do. I wanted to go back to working in film. Even better, I wanted to direct documentary films. So, I decided that to go on this trip, I needed to go as a filmmaker. To do that, I needed to become a professional in my craft, but at this point in my life, I was not re-enrolling in an institution of higher learning. I didn't want the conformity or the debt associated with it. So, instead, I decided that to subsidize my education, I would sidestep the system and enroll in the school of hard knocks. I was going to be self-taught, and this trip was going to be my first course in the program.

What I didn't realize at the time was that this self-sufficient choice to create the program, define the terms, and appoint the credentials was formative and relevant to everything I would become. It was a crossing-over point in my development. I was now a singular seed germinating into an institution of self. I was no longer looking outward for systems to validate me and my goals. Instead, I turned inward with the belief that at every point in life, one can choose their path, define the course objectives, and become something that the institutionalized world argues is not possible to do without their permission. Gatekeepers be damned! I was on a course of self-determination.

After several days of thought, I was confident in my new direction. Picking up the phone, I called Dan back and shared my idea. I told him that my objective for the trip was to film the experience and make a documentary. For him, it was an excellent plan—now he had his first crew member, and things were moving forward.

Within a week, I drove up to Alameda, California, where the boat was parked in a slip. There, in preparation for filming our

CHAPTER 6

maiden voyage, I pulled out my Sony Hi-8 Handycam my parents had given me for Christmas just a couple of years prior. Rolling tape for the first time, I began recording the first voyage on the San Francisco Bay, where we were silly and slap-happy on camera. As unskilled as I was, there was something new, fresh, and novel about the footage. It had that rare unscripted look that documentary filmmakers love. Although I had no clue that what I was doing, that day would capture the opening scenes of the film and set the stage for everything we'd do over the next twelve months. I kept rolling, and Dan kept talking.

After a few trips and a little time doing research, I realized I was a long way from being a director or professional behind the camera. In addition, after reviewing the Handycam footage, I could tell that the quality wasn't there for a professional piece. It was low resolution and grainy. I needed to invest in some new equipment and take my first step toward a real commitment.

It was still pre-YouTube tutorials and product reviews long before Amazon was a powerhouse in online retail. So the work of finding the "right camera" depended on talking to the right people. I was living off the grid of the industry, so to speak. Fortunately, Santa Cruz, a surf town, did have its share of surf filmmakers. So, I started looking on Craigslist to find some resources in the area.

Around the same time, I had started to go to a small theater on the north end of town that screened independent films. On weekends, while I had nothing else to do, I would go out to see documentaries that were not mainstream. It was a cozy theater with red velvet seats. I loved going in with a bag of popcorn, some Junior Mints, and a Coke. Sitting down center stage, I would watch independent documentary films like *The King of Kong, Who Killed the Electric Car, Waltz with Bashir,* and *The Bridge.*

One weekend, after I'd committed to filming the trip, I attended the screening of a surf film called *Unsalted: A Great Lakes Experience.* As the title suggests, it's a film about surfing in the Great Lakes and the experience of chasing waves in freshwater. The piece was outside the norms of surf films, which were usually filled with pro surfers catching exciting "rad waves" on dangerous, powerful, epic breaks like *Pipeline* or *Chopu*. The film *Unsalted* was more about the kooks of surfing or the goofy-quirky unskilled people who do it for the love of it and its euphoric, thrilling "stoke" that transcends skill or location. I loved it.

Improving on the experience, the film's executive producer, an ex-pro surfer, was there. His name was Ian "Kanga" Cairns. He was a member of the Golden Aussies, a rare group of individuals from Australia who successfully broke into the nepotistic surf culture in Hawaii. Doing so, however, came at a cost, and he and the crew had to flee for their lives from the crazed locals after winning the national competition. In doing so, he'd gone down as a legend in surf history. To see him firsthand was a thrill.

With credits finishing up on his film and people applauding, Ian hung around afterward to participate in a Q&A session on the film. After a few minutes, the Q&A wrapped up, and I wanted to speak to him personally. Seeing an opportunity, I jumped in line behind a handful of others and waited my turn to chat.

When it came time, I introduced myself, expressed my appreciation of his work, and asked if he was looking for anything new. When he said "always," I saw my opportunity and quickly promoted my idea for a surf film of guys on a boat. To my surprise, he liked it and offered me some tips on cameras and equipment for the sea. Then he said, "Once you have a script and outline, send it over to me, and I'll take a look."

Ecstatic, I left the theater feeling like I was walking on cloud nine. I now had an assignment from a professional and the basics I needed to graduate from amateur and move to novice status.

Within a week, I'd purchased my new camera, a $5000 Sony FX1, and was working nonstop to write my script. I wanted the story to be about more than the self-absorbed reality of filming ourselves in a boat going down the coast. While that was part of it, what I wanted more was to capture the essence of the experiment that I was running on myself. I wanted to tell a story about what happens to a dream when it becomes a reality.

Traveling north on the weekends, I began to capture small moments and scenes while we prepared the small boat for the likes of the Northern Pacific. Over the next nine months, we strengthened the gangway entrance on the boat, built a storm hood, drilled scuppers to drain off water, and reinforced the boat to protect and prepare for worst-case scenarios of waves that could sink us.

In between working on the boat, there were humorous behind-the-scenes rants on camera depicting our suicidal courage or incompetence, arguments about who was carrying the most workload, and intimate moments on deck fighting over who would eat the last McDonald's fries. Finally, after nine months of work and preparation, I had captured a slew of footage, and we were ready to go.

In the days before departure, I resigned from my teaching position, moved out of my apartment, and broke up with the girl I was seeing (or rather, she broke up with me). Again, homeless and alone, I packed up my life's belongings, stashed them in my car, and parked it at a friend's farmhouse. Pausing for a moment, I pondered what the transition would be like when I returned. What I hadn't figured out was the challenges that life on the sea would present before leaving to go. I was soon to jump off a cliff into the deep waters of the Pacific without a clue what I was getting myself into.

DOWN BAJA PENINSULA—AT SEA WITHOUT CREATURE COMFORTS

To embark on our journey, I had narrowed my life down to a duffle bag of clothes, a sleeping bag, and my camera kit. On February 23, 2006, in the dead of winter, Dan pulled out of the Alameda Marina and sailed across the San Francisco Bay under the Golden Gate Bridge and into the mighty Pacific. We had officially departed south on the scenic California coastline, heading down Big Sur and Point Conception. I was now on a trip that needed to be captured on screen, not just on paper.

In my opening for this book, I spoke about a moment at Taylors Falls in Minnesota when I climbed to the top of a cliff and, with little to no preparation, jumped off. That story was a metaphor for how I had lived much of my life up to that point and beyond. And despite having learned my share of lessons, I was still jumping off cliffs nearly a decade later, unprepared for the landing. This trip was no exception.

After two weeks, I was beginning to come to terms with all I'd given up to make this trip happen and what I was getting in exchange. In this time, I'd found it hard to capture on camera any great moments. The boat was always moving or rocking and creating shake in my shots. The movement made it almost impossible to film with a steady hand. Furthermore, the scenery was often the same: the coastline in the distance, the ocean up close, and the two of us—Dan and Barry inside the boat talking. I had thought this would be like the original road trip via Jeep on land, but it was nothing like that, and it felt like a recipe to flop. My grandiose dreams of fame rapidly deflated, along with my ego.

Adding to this experience, living on the boat was far more harsh than I'd expected. It was winter in California—cold and rainy most days. I began to yearn for some creature comforts inside, but there was no getaway or escape from being cramped up in the small space of the boat. In reality, I had nowhere to go. I was a trapped animal wanting to get out. When I wasn't obsessing over all I had given up for this trip or taking out my frustration on Dan, I was just sitting there brooding and increasingly filled with regret.

En route south, having passed Big Sur and Point Conception, some of the most treacherous seas on the Pacific, we arrived off the coast of the Santa Cruz Islands west of Santa Barbara. Anchored in a small cove called Potato Harbor, the sun had set for the day at around 6:00 p.m. Tucked inside the boat for the night, it was dark out, and we were trying to stay warm. Dan had fired up a Coleman lantern, and with lots of time on our hands and nowhere to go, he'd picked up my camera and, for the first time since leaving, pointed it at me. During the past two weeks since departing, he had seen the highs and lows of my emotions, watched the obsession of my thoughts, had been worn down by

CHAPTER 6

it all, and wanted some answers. Grilling me for several minutes, he got down to the bone of the problem and was cutting in deep.

"Bowski, you seem like this trip isn't going well for you; if you're unhappy and don't want to be here, why stay?"

Hearing this, I was taken aback. I began to retract because I didn't actually want to answer these questions. I preferred to just be disgruntled.

"Seriously," he went on, "Dennis gets on the boat in a few days. If you aren't enjoying yourself, why not just leave?"

His questions struck a nerve in me. The truth was I wanted to abandon everything. At this point, the idea of becoming a director of my own film seemed dumb. Building my own institution of learning through hard knocks felt stupid. Deep inside, the little boy was again flush with shame. All the trauma of the third grade came crashing back. Inside, I could hear a voice say over and over again, "You are a failure and have always been a failure, and you will fail again, so you should just give up now."

Ironically, I wanted to run away into nature, but nature was all around. With nowhere to run and unable to answer Dan's question, I just said, "Right now, I am not ready for that. I am not ready to quit."

Then, I retired to my bunk for the night, crawled into the fetal position, and shut down.

The birth in the front of the twenty-four-foot boat was somewhat like a coffin—traditionally called the "V-birth" from its v-like shape caused by the design of the forward aft portion of the craft. The bed was snug. While at my head, there was a solid five feet of width from side to side at my feet, it was merely eighteen inches, and the length from head to toe was barely long enough to fit me. Because of my height, I was forced to sleep at an angle with my knees slightly bent. On my back, my nose was about an arm's length away from the ceiling, and if I tried to sit up, I would smack my head on fiberglass or worse (pin it) on a "through-bolt" sticking out from the ceiling.

At night, the humidity inside the cabin from our breath and bodies would collect into beads of water that would drip down onto my face like Chinese water torture. On this night, the drips smacked me square between the eyes and fully woke me from my slumber at 3:00 a.m. There, alone in the darkness and loneliness of the craft, my stomach churned in a ball of regret.

In my bravado before departure, I believed I could define my own path. However, I was realizing that the road I'd chosen was far more difficult than I'd ever thought. In fact, I began to think it was impossible. But I was soft and did not have the internal fortitude, experience, or mentality needed to succeed. In reality, I was merely a shell of what I had claimed to be; my words had only been words, and this boating life experience was fleshing out my weakness, and I was fighting it. As I lay there with drips of sweat rolling down in the dark, I felt stuck. I had followed the famed slogan echoed by many motivational speakers: the words of Spanish conquistador Hernán Cortés, who decreed, "Burn the boats and take the island." I took the advice and, like his fellow sailors, could not go home, but full of fear, I did not want to go forward. Stuck, I was caught in indecision and needed a resolution quickly.

KEEP TAKING RISKS—THE FIRST LESSONS OF SUCCESS

Over the next few days, it was all I could do to find the courage and strength to pick up the camera to film. Emotionally, it seemed pointless, and I had fallen apart on multiple occasions and could be heard off-camera yelling profanities at myself in fits of rage. In desperation, I decided to take a risk. While anchoring outside a spot called Lefts and Rights at The Ranch north of the town of Gaviota, I packed my camera into the waterproof Pelican case, stowed the tripod inside the rubber dingy, and paddled to shore.

Thrilled, I made it past the surf and was safely on the beach. I could now put up a tripod and compose a shot. Walking around, I spent the day capturing sea life in the tide pools, steady shots of waves peeling along the reef, wide pans of the California coast, and beauty shots of nature that I'd desperately wanted to catch. As the day passed, and having finally made some progress on the

CHAPTER 6

film, I packed everything up on the dinghy and began to make my way back out to the boat.

In the time that I'd been on shore, the tide had started to come in. Unbeknownst to me, the surf had picked up, and I was now concerned about getting out. Watching for lulls, I worked to time my escape back to the boat, then launched the dingy and started rowing hard. I could feel smaller waves pass under me as I worked to row my way back to the boat. Getting past the initial foaming white water crashing onto the beach, I'd assumed I was in the clear as I made the final push beyond the surf toward the boat. Then, like the famous scene in the film *Castaway* with Tom Hanks, a swell popped up on the horizon. Catching it out of the corner of my eye an outside set, I knew this was going to be trouble and began paddling for all my life.

Dan, who'd been watching from shore, had begun to swim out on his surfboard and was not far away when catastrophe struck. Looking over my shoulder one last time, I watched the wave double in size as it hit the reef and began to come down on top of me. Without a moment to think, the entire craft with all its contents inside was picked up, tossed upside down, and thrown backward into the surf. Popping up above the water, I screamed, "My camera!"

Scrambling, I began to dive into the water in search of my Pelican case, which contained the entire purpose for being on the trip. Losing this in the surf would mean the end of the trip as a whole and everything I'd worked at for nearly a year. Wave after wave passed, and nothing came up. As the chill of the Pacific began to set in, I felt certain that everything was lost.

Nearing tears and defeat, Dan raced over to help. Flipping the dinghy over to right the ship and get back on course, he was doing what came instinctively. There, to my surprise, floating under the little boat was the Pelican case on top of the water. Overwhelmed with relief, we still had to get it back to the sailboat and see if it was dry. Grabbing it, I immediately threw it in the boat and, with Dan's help, we worked to get past the breakers and outside onto the boat.

Drying off the outside with a beach towel on deck, my heart was racing. I paused before popping things open. At that moment, I knew that if I looked inside and the camera was destroyed, the trip was over, and I was going home. But if not, the trip was saved, and for better or worse, I was staying here. In a twisted way, this moment was making the decision that I had been unable to make myself. Unclipping the latches, I lifted it open. There in front of me, surrounded by protective foam, was my $5,000 camera, dry to the bone! I couldn't believe it. It had been protected by the case in the raging surf and didn't have a drip of water on it.

Over the next few weeks, I had little to offer or say. I had been such a coward drenched in my own self-pity, humbled, and broken down by the sea that I felt like there was just nothing to contribute. But I knew the important thing was that I was still here; I had not quit.

Starting over, metaphorically, I began now to rebuild from the inside out. Pushing the record button and looking at the frame, I began to find good moments to capture on film and scenes that I could piece together into a story, and I could feel it all coming together one piece at a time.

Traveling south from the Santa Cruz Islands, we passed through my old stomping grounds of Marina del Ray. There, on the coast of LA, Dennis joined the trip, so my responsibilities assisting with the sails, anchoring, food prep, and cleaning were reduced. After his arrival, we journeyed to Catalina Island twenty-three miles off the coast of Southern California. Then, we headed south toward Ensenada, Mexico, to sail for eighteen hours overnight, a miserable experience of sleepless, damp, cold, throbbing work. There was no autopilot, and we steered the boat by tiller (or a long stick in the cockpit used for leverage when turning the rudder). On these nights, someone needed to be awake throughout. Pulling shifts, we struggled and suffered nonstop until we reached the marina and could shut down.

Once departing Ensenada south, we'd be entering the most barren portion of the trip. For the next 775 nautical miles, there would be no town with a marina in which to recover, a first since leaving San Francisco. Prior to this, we would go out to sea and

CHAPTER 6

anchor at islands along the coast. Exhausted from the exposure to the open ocean and being worked over by the nonstop motion of the seas, we'd head to a nearby marina off the coast of California to plug into electricity, take showers, and escape in recovery from the rocking, wobbling motion of the sea. However, after leaving the marina at Ensenada, there would not be another place along the barren Baja coast to recover until we reached the tip in Cabo San Lucas.

In preparing to depart, we decided to slowly double-check all supplies and relax in the sun a little longer, knowing things were going to get harder, dirtier, and leaner from here on out.

Sitting in a chair on the deck of the marina, I watched as Dan repaired the sails high above while playing cribbage with Dennis dockside in the sun. Here, I pondered how our last few weeks had passed since those dark, cold days off the coast of Santa Cruz Island, where Dan had interrogated me and the tragic dinghy accident on the shoreline of California. I was starting to feel proud of myself; I had survived my storm of emotion and defeat, had crossed over what felt like an abyss of my fears of failure, and was feeling stronger and more confident every day. The script that I had written in preparation for the film before leaving was now returning to my memory. Ironically, I had predicted the painful struggles of our early days. In recounting the story in my head, I recalled posing the thematic question, "What happens when a dream becomes a reality?" I wasn't sure at this point, but I knew that the dream, as it stood, was lost, and our reality was no dream.

As we lingered, I recalled in retrospect how, at the start of the trip, the dream was picturesque. In my mind's eye, our trip would be like a catalog of the world's best sailing magazine. Like a projector on a movie screen, I saw days spent sailing slowly over blue waters under blue skies. We'd anchor in emerald seas with palm trees and a light breeze. On afternoons, we'd surf on warm waves, all to ourselves. And during nights on board, we'd talk together, eat fish we'd caught, and play board games under the boat's golden lanterns. It was truly pure fiction, the stuff that advertisers love, but the reality could not have been further from the truth—the truth was that the trip entailed far more suffering than pleasure. However, when pleasure did come, it was acute in ways that could not be experienced otherwise.

CROSSING OVER: THE DREAM BECOMES THE REALITY

Compared to the routine life in California, crossing the sea was hard daily work. Being on the boat worked every muscle in your body to maintain balance. In those early days, our physical weaknesses brought on sea sickness. Sailing day in and day out chapped your skin, and the pulling of lines wore at the fingers and hands. The rubber wet-weather gear, a must while sailing in the colder north, got clammy and stunk like sweaty armpits and butt cracks. It's an aroma that you could neither clean out nor escape.

Leaving Catalina Island some weeks prior, we'd work the boat in the twenty-foot seas. Waves of cold water came splashing over the deck, drenching the cabin and cockpit. Wrapped up for warmth, you could feel our pasty gear clinging to our skin, and we'd begged for warmer days where the warmer weather would lend to no gear. On the water in the early months, the temperature was always ten to twenty degrees cooler than on land. So, all in all, you seldom felt comfortable.

At night, if you were lucky, you'd pull into a calm anchorage and rest, but if you were unlucky, you wouldn't. Settling into a spot on a rough night, you'd drop two anchors, one to the aft and one to the bow. The aim was to get the boat facing into the oncoming seas so you got more pitch, front-to-back, which was slower and more controlled, versus the roll of the boat going side to side, which would exhaust you and toss everything around through the night. On those nights, it was the worst. Dan stayed awake watching so that by the morning, an anchor wouldn't drag or come loose, leaving us on shore, or worse, crushed up against the rocks, doomed to sink.

CHAPTER 6

Pressing my head into my pillow on rough nights, the boat would float back to forward in jarring motions. Each time, the wet and tethered ropes would pull taut against the fiberglass hull and create a popping and creaking noise that echoed louder and louder inside the cabin, making it nearly impossible to sleep.

On any given night in those first days, you had to contend with the aforementioned drips of humidity and sweat, crackling anchor lines from hell, or the smell of one of your comrades waking to pee in the middle of the night into a small piss jar, mere inches from your head. This was designed to avoid having to go above deck and hang yourself overboard, a dangerous stunt that late at night, but it was unattractive, nonetheless. If you were fortunate enough to sleep through all of this, you were certain to awaken at the morning dew point, soaking wet in your sleeping bag and rummaging around for a dry shirt and underwear to recover. The torture of those early days was not something I'd wish on my worst enemy. However, while slow as a snail, time did pass until, finally, having crossed deeper into Mexico by mid-March, the air and sea started to warm, and our bodies began to grow stronger, making things more bearable to the soul.

With waves now slapping the side of our boat, we were back out to sea. Sixteen kilometers in front of us were two small islands, the names of which were "All Saints Norte" and "All Saints Sur" (more appropriately known in Spanish as "Isla Todos Santos"). Reviewing the land from a map, you can see that the collective islands make a straight line extending at a forty-five-degree angle pointing from northwest to southwest. Collectively, they make up one piece of land, with a shallow isthmus of emerald-green water separating them in the middle. That isthmus layered with car-sized boulders is full of orange Garibaldi (the official protected state marine fish of California), lobsters, starfish, and a host of amazing creatures you can see swimming through the reef. Having made the passage from land, we were slowly traversing between the islands, like a spaceship floating through the void of the universe.

Over the southern island, there was a ridge of sharp, jagged peaks with cliff faces and ridges jutting up and down, making it uninhabitable and nearly impossible to climb. In contrast, the northern island was a plateau easy to walk across. There was a stairwell up one side of the cliff face, and you could see a black and white barber-shop-looking lighthouse on the furthest northwestern point. On a clear day, this towering figure could be seen from as far away as the mainland in Ensenada and marked the location of an amazing wave.

Jutting out from this island past the lighthouse was a natural jetty wall creating a perfect break for surf. At that spot, there was a famous big wave break called *Killers*. Breaking over the rocky bolder reef bottom, this spot was not to be contended with. The takeoff was fast, and the reef was dangerous, but we were traveling out here to take it on.

Killers had reached its peak of notoriety on February 16, 1998, when Taylor Knox, a big wave surfer from California, made history. In that year, he won the K2 Big Wave Challenge, a competition to catch the biggest wave on breaks around the globe. This victory put *Killers* on the map after he was photographed taking off on a glassy fifty-two-foot blue avalanche of water. On that day, surfers, jet skiers, photographers, and journalists were around to capture the experience that ended up on the cover of surf magazines. However, when we arrived, the place was empty with not a single person around. We had it all to ourselves and started to pull into sight; we'd hoped to find it big enough to break and small enough to be manageable.

As we crossed the passage between the two islands and entered the other side, we got our first look at *Killers*. Our weather report had said that the swell was only six to eight feet—rather small for this spot. So, in our passage over, we were uncertain that we'd see anything of significance. However, at first glance, we were elated. The tide was building from its morning low, and the reef was setting up a nice peak that was pitching over a takeoff with a slower wave shoulder—one that ran a solid fifty yards or more before hitting the rocky shore.

CHAPTER 6

Never one to waste time, when there was surf around to ride, Dan suited up, jumped off the boat, and started his swim to the lineup. While Dennis waited onboard to measure the safety and feasibility of the ride, I wrapped my camera in a dry bag to protect it from water, then hopped on a board and swam over close enough to get a shot. On camera, you can see Dan spotting a wave and starting to paddle for a shoulder. In seconds, the wave shaped up as a barrel doubling in size upon hitting the reef below and throwing water off its lip. Dan, sitting just off the center, paddled hard. In an instant, the wave picked him up and shot him down the lineup. For several seconds, I watched on camera as he ripped powerful, aggressive turns from top to bottom. From fifty yards away, I could see the spray coming off the lip. It was a signal from behind the wave that he was still on it. Then, just as the wave started to close out near the rocky shore, Dan turned sharply, pulling out, then let out a classic yelp that echoed off the surrounding rock walls of the island. We had officially arrived.

Dennis then joined, and for nearly an hour, I worked to film while floating in the water off my board. This was some good surf footage. I was no longer getting beat down by frustration. The film began to have legs, and my confidence was on the rise. For the first time since the trip began, I was no longer dreaming about making a great surf-adventure film. I was actually doing it.

THE WHALE, THE WAVE, AND THE MESSENGER

As I captured moments while floating above the crystal-clear water, I could look down and see the rocky bottom of the ocean some forty or more feet below. The emerald-green waters were glassy, the wind was calm, and the visibility was high. Then, to my complete surprise and dismay, I spotted an object moving that was much larger than a dolphin just beneath me. With my brain trying to make sense of the experience, I watched as a giant tear-drop-shaped baby gray whale swam some twenty feet beneath me.

Startled, I couldn't believe my eyes. It seemed so unreal, even impossible. Quickly, I took a second glance to be sure I wasn't imagining things. Looking down at the gray object, it looked big enough to be a submarine. It even cast a shadow on the ocean floor below as it passed. Frozen in place, I had never in my life been more in awe and scared at the same time. Unable to contain myself, I yelled out to Dan and Dennis, who could not hear my cries because of the roaring, breaking sea.

Then, the small giant startled by me (this little creature floating above), kicked its tail, and quickly disappeared into the distance. Watching, I could see the powerful currents boiling up from the sea in its path. Exhaling a breath, I was overwhelmed with the sense that I had been visited by a near spiritual being. The experience was so rare and unique that it inspired me to question its meaning and purpose. Why had it come to visit me? What did it mean?

My mind flashed back to a few years earlier when I drove down the Baja and across the country by truck. During that trip, I was joined for a portion by a young lady-friend named Leslie, who had taught me something in Wyoming with which I had deeply connected. Leslie, an advanced traveler and adventurer in her own right had at one time in her life lived and taught for several years on an American Indian reservation out there. During her time there, she'd embedded herself deeply within the culture of the tribe. In doing so, she experienced ancient ceremonies and a sweat-lodge fire meditation that few white people ever experience. There, among the tribal members, she learned and came to believe that all animals were spiritual creatures and a direct encounter like what I'd just experienced was transmitting a sacred message. I just needed to interpret it.

That night, back on the boat, I broke away from the crew and sat alone above the deck to think. For the first time during the trip, I decided to record some direct commentary on camera. Sitting there, I began to open up about how I'd missed so much time onboard the boat being focused on all the wrong things. Having a moment of clarity, I realized that I had been lost, deep in fear, and that fear had driven me to want to escape. Over and

CHAPTER 6

over again, I had been obsessing about the past and what I'd left behind. That obsession had sent me deep into depression. But having been visited by this giant baby whale, I was pulled crisply into the present, and all my depression faded away.

There below deck, laughing, talking, and reading books were my two "compadres." This and every moment with them on the trip was rare and unique. I realized with already half the trip having passed, that these were adventures that I had never experienced before and would never relive again. Fully present in the moment, I let out a big sigh and began to accept reality for what it was. In that breath, it became clear to me what the transition between a dream and reality truly is. Turning on the camera, I started rolling and began to narrate my thoughts, "When a dream becomes a reality, it is no longer a dream. Because a dream is fiction, and reality is the here and now."

I was striving at the moment to articulate that a dream is only a fleeting idea that remains in a state of suspended thought without action. On the other hand, action is what is happening in the present and is physically no longer a dream. Hence, the dream and the reality operate on two different but parallel planes of existence. Therefore, the only way to transform dreams into reality is to be present and take action daily against all struggles without ceasing. In doing this, the transition between the two can and will happen, and you can experience the dream in the most real sense of the word.

Having had that clarity, I cut off the camera and returned below deck to be with my friends. From that point forward, the reality of the trip began to merge with the dream, and the dream started to become reality. The drama and problems that once haunted me day in and day out no longer felt important. Material things and money felt secondary. What was primary was enjoying the process. From my confusion came clarity, and clarity brought faith. That faith was the acceptance that it was all going to work out in the end, and my role in that process was to stay present and do the work. It was the first time in my life that I had actualized my belief.

Over the next few weeks after departing Todos Santos we traversed slowly south, anchoring off the coast of the Baja and its off shore islands. In the protected coves of the uninhabited shores, we stopped at consecutive islands: Isla San Jeromino, Isla San Martin, Isla Natividad, and Isla Cedros. Each was a variety of expired volcanos that left behind cooled molten rock or deposits of dirt jutting above the ground from the mountain range below water. Some were sandy breeding grounds for pods of seals. Others were bird refuges full of wildlife uninterrupted by human development. Walking around each, I noted that some islands were so full of birds that you would think birds ruled the earth while other islands were so barren that it seemed the apocalypse had come and we were the only ones left.

As we sailed farther south, we uncovered fishing villages on Isla Natividad and Magdalena Bay that seemed forever suspended in primitive times. On Isla Cedros, we discovered a salt mining operation with giant cargo ships that tapped into the global economies. At Belcher Cove, we uncovered shark fishermen living in isolation rolling out thirty miles to sea every day to make a living selling all the remains of mako sharks at market. On Isla Cedros, there was a solo resident feeding his family on a state-funded income living there; he'd stay two weeks on and two weeks off to maintain the lighthouse.

Navigating through each place was like traveling through space to planets never before discovered. Mentally, it was surreal; physically, it felt fictional; and holistically, the experience was beyond rational explanation.

Onboard, things had settled into place as well. Dan and I had had some drama but nothing compared to the road trip, and overall, we were in a good spot. Dennis had started to open up and show more of himself too. And the three of us had become a sort of tribe. Out here, we had been cut off from all forms of contact with the outer world in our entry deep into the Baja. Down here, there was no email, no phone, and no way to mail anyone a note. For weeks on end, both day and night, it was only the three of us and the sea. This was a sort of phenomenon of isolation in

CHAPTER 6

our age that we'd never experienced before and became a catalyst that inspired deep bonds that remain unique to this day.

Out here, it was only us and us alone, and we no longer needed words to communicate. In this space, there was texture between us, a common loyalty to the group, and an awareness of all our weaknesses and strengths that were no longer ours alone.

In our own way, we began to mirror the culture of the small fishing villages we'd visited peppered along the coast in pockets and places unknown by most of the world. Here, the people lived in what would be considered complete poverty by global authorities, but what I saw was a connection with the environment, as they lived in houses made of brick and shacks built from materials off the land. People fished the sea daily for their sustenance and wasted none of what they harvested.

With their lean figures and tough skin, you could see the people had been hardened by the land and were pure of spirit. They were willing to share goods in exchange for company. To our good fortune, we ate homemade tortillas on a homemade stove and fish fried on an open pan fire in exchange for some time to socialize and share our story. You could tell we were among the people of the earth, and in a way, I felt a connection to my childhood fish fries out here. There were no classes, groups, or hierarchy, and the human experience, not money, was the highest form of exchange.

Throughout our exploration, we maintained the goal of uncovering un-surfed waves and naming them as our own. Time after time, we circled the islands in search of these waves, hoping to uncover our spot. Finally, while circling the island called Isla San Martin, we saw a wave that I was certain had never been surfed. Watching from the cliff face above, I named it "Places Unknown." But this tragically remained unofficial as we had no boards in our hike and never got back to paddle out past the jagged rocks and boils and ride a wave.

Having searched coves and breaks for two months to capture that "perfect wave," we finally found it, but it had already been discovered years before our arrival. It happened after an overnight sail from Isla Cedros to the sheltered cove of San Juanico.

Having arrived at 3:00 a.m. and exhausted, we crashed into our bunks for the night, hoping for something amazing the next day.

In the morning, we poked our heads out, and there at the end of a large hook-shaped bay along the coast was the famous long-board wave called "Scorpion Bay." Cooking up some pancakes after the long eighteen-hour transit the night before, we could feel the swell pushing under the boat. Dan, ever the pioneer, could not wait any longer and soon left us behind in the comforts of the boat to paddle over. Making it to shore with my camera, I could see the lines of waves coming in from the Pacific. Standing there, I watched them peel like a machine, one after another. Never before in my life had I seen a wave break with such precision. Over and over again, it unfolded from the first point down to the beach for well over a quarter mile.

Popping the camera up on a tripod, I turned it over to Dennis to film. He then focused the lens on Dan and rolled. Watching, we could see as Dan swam hard into a nicely formed four-foot wave and took off. Quickly, he popped up and started working his way down its unbroken face, pumping his board rhythmically up and down over and over again. On and on, he rode it for what seemed like an eternity, catching easily the longest wave of the trip and possibly our lives.

As I stood behind Dennis and watched the experience, it felt like a scene from the film, *Endless Summer*—a film that had inspired much of this lifestyle for me. I had first seen the film at the age of fourteen while living in the Midwest. It inspired me then like no other film before it. Out here, we were experiencing our own version of the *Endless Summer,* and I was living out one of my dreams in real-time. As Dennis kept rolling the footage, I continued watching until Dan, in complete exhaustion, dropped off the wave into the white water below. Reviewing the footage in the documentary some years later, I timed the ride. From end to end, Dan rode for well over a minute, which is rare for any surfer in the sport. It was a highlight of the trip and the film the "Perfect Wave" and one that I still cherish watching on camera to this day.

On the final day of the trip, we reached the tip of the Baja and sailed into Cabo San Lucas. Entering the bay, we were suddenly

CHAPTER 6

surrounded by cruise ships, jet skis, and beach-going tourists of all sorts. Coming from two months of near-complete isolation and survival, I now felt painfully disconnected. This culture that I was already conflicted by now seemed completely foreign to me. Having embedded ourselves so deeply into nature and seen fishing villages connected to the land, it became clear that these cruise ships, hotels, tourists, and condos were void of a connection with this place. They were transient at best, only taking for themselves and depositing little in return.

Over the next few days and weeks, I'd again go through the shock of reentering my own culture. Part of me would reject it, never fully accepting the terms of returning to the bliss of the ignorance I'd once shared. That part of me would always feel a little lost long beyond the trips and the traveling would come to an end.

Off the boat, we found a hotel and took a few days enjoying Cabo before the three of us prepared to fly back to our respective lives. Dan would go on to medical school to become a doctor. Dennis would return to teaching kids in Los Angeles, and I was returning to San Francisco homeless, needing to rebuild my life and get a job. While I didn't have much to return to, I did have a box of tapes filled with over one hundred hours of footage. That footage possessed the potential to fulfill a dream of making my own documentary—a dream that would take another fifteen years before it would see the silver screen.

Barry shooting for Animal Planets, Weird, True and Freaky season 4, eps 2 titled "Life on the Dangerous Edge", taking a break away from set to see glaciers.

"Go alone, go fast, go together, go far."

—African Proverb

CHAPTER 7

NATIONAL TELEVISION AND THE FALL (2007 - 2009)

THE DREAM JOB

It had been two years since the filming and completion of the sailing trip. In that time, I had gone from being nearly homeless and sleeping on friends' couches to being employed and living in an apartment in the Inner Sunset, a neighborhood of San Francisco. In a twist of good fortune, I was hired as an associate producer on a show for the *NatGeo* TV network called *Defying Niagara*. The job was a dream come true. My work on the sailing documentary arguably clinched the deal. As I integrated back into society and transitioned into living in San Francisco, I now had two of the three things I desperately needed: a place to live and a job. The only thing missing now was love.

Since the chaotic life on the boat, I started to rebuild my routines, hobbies, habits, and all the little day-to-day things that are taken for granted by the average person. In the morning, I would get up at 6:00 a.m., shower, grab some coffee, and drive to work forty minutes away in Emeryville, California. Production work was, is, and will always be a slog, so I seldom got home until 7:00 or 8:00 p.m. But because I did not like to be alone, I would then go work out, grab some food, call friends, watch TV, and finally shut down by midnight before waking up the next day and doing it all again.

As busy as I was Monday to Friday, on weekends, I had time to surf. If it was local, I was paddling out at Ocean Beach (aka OB), Deadmans, or Fort Point. But what I really liked was a day drive down the coast toward Santa Cruz checking out Cracks, Davenport Landing, Three Mile, or my favorite secret little wave, Rims. However, more often than not, because of all my moving from place to place over the years, I usually was doing it alone. Thinking back, I couldn't help but feel echoes of regret from the three big moves I had made in those years. In a relatively short time, I'd moved from college in Minnesota to Los Angeles, from Los Angeles to Santa Cruz, and from Santa Cruz to San Francisco—four major cities, three major road trips, and a long boat ride in just over five years. It's amazing to think about it. Each time I moved, however, I was rebuilding my life, and doing so began to wear on me. I was dried up and felt more transient than ever. I was neither a local nor a tourist, a friend nor foe. I was a vagabond lost in the backdraft of my own making.

To cope, I worked all day, every day. During the week, it was *NatGeo* for MHP Productions. During the weekend, I would edit on personal projects. I also started filming a new documentary on UltraRunner Mark Gilligan, who was training in the jungles of Hawaii for a one hundred-mile race called *HURT100*. An UltraRunner is anyone who runs races longer than a marathon's distance of 26.2 miles. To capture this, I had to think of creative ways to keep up with Mark as he ran in the sprawling redwood hillsides outside of Oakland or on the coastline outside of Point Reys. I would study the trails for shortcuts to where he would reappear in the loop or race around in a car to catch him at various checkpoints. It wasn't easy to do, but the trails he ran were beautiful and made for great footage as he covered the insane distance. Each training run was a minimum of twenty-five miles. A marathon was a warm-up run for the races this guy ran.

CHAPTER 7

On one occasion, I met him at his office in downtown San Francisco. A programmer, he was working on the thirty-fourth floor of a fifty-two-floor skyscraper coding. Spending his days there—punching on keys locked inside the cement jungle, he needed an outlet. To keep up with his regimen, he'd take his lunch break and run the fifty-two flights of stairs in the fire escape stairwell to the bottom multiple times. Carrying my camera rolling, I would do my best to keep up with him. Most often, I would fall behind, then cross paths again as he looped around, heading back up from the bottom. It was some of the craziest stuff I'd ever seen.

When I wasn't working doing TV or filming Mark, I was digitizing the footage of the sailing documentary (transferring it from tape to hard drive). With over one hundred hours of footage, at one hour per tape, it was a full-time job to capture and log, which took months on end. Now my life had boiled down to doing one thing: work, work, and more work. It was all I had, and I vowed not to stop till I reached the top.

THE DAREDEVILS WHO RISKED IT ALL

For my first national television production, I was working on a show called *Defying Niagara*. It was a show documenting the history of daredevils who had taken barrels over Niagara Falls. It was a sixty-minute documentary for which we had a budget of around $350,000 and a production schedule that lasted around nine months. My job during that time was to dig up every daredevil who had gone over the falls and was still living. Once I had found them, I needed to organize an interview with the show's producer and collect every piece of archive from the event they possessed. Basically, I was a research journalist digging up some of the most unique and eclectic people on earth, and I loved it.

The first thing I needed to do was hire a private investigator to get the personal info of each daredevil. Having never worked with a PI, I asked around the Bay Area for a recommendation, and I found one living in Chinatown who came highly recommended. I hired him, then waited for his callback. Within a few days, an email arrived and, with it, a long list of names and addresses: Dave Munday, Steve Trotter, Geoffrey Petkovitch, Peter DiBernardi—all barrel riders, and Jessie Sharp, who'd made a solo canoe attempt. The investment paid off. I had my subjects' names and a long list of potential phone numbers and was off to the races making calls.

One by one, I started calling the stuntmen. Working through the list of numbers, I would, at times, get a laugh from people with the same name but not the same experience. Working my way down the list, I finally got my first stuntman, Steve Trotter. In 1985, Steve made his first successful attempt at the age of twenty-two, which made him the youngest person to ever go over the falls. After championing the stunt once, he decided to go for a second time in 1995 at the age of thirty-two in tandem with the first female to ever successfully go over the falls.

When I finally contacted him, just over ten years had passed since his second successful attempt. On the other end of the phone down in Florida, I could hear his raspy-worn voice and could tell the years had been tough on him. He was the type of person who was running on different fuel than the rest of us. Now in his fifties, he sounded tired from the ride but as excited as ever to hear from us.

Ever an opportunist and wanting to maximize his position in the show, he started to pitch to me that he was planning to go over a third time. For me, it was simple; I wanted nothing to do with encouraging such a stunt. Side-stepping his interest, I asked if we could come down to the beach where he was living and interview him. He agreed. And I had my first daredevil for the show.

Feeling more confident in the process, I picked up the phone to call Jeffrey Petkovitch, who quickly became a favorite for the show. Jeffrey had originally joined as a bet with friends to go and was headed up by Peter DiBernaldi, a used car salesman we never found. Peter seemed to be using the stunt to increase the visibility of his company. After weeks and months of work, he recruited

Jeffrey, a local youngster who was crazy enough to agree, and the two went on to Niagara Falls infamy in September 1989.

By now, a husband and father of two young kids, Jeffrey was still wild but more contained and could laugh about the madness of the event. He said, "The barrel was a fortress built with oxygen tanks and radio communication that recorded the audio of the experience verbatim." Then, warming up to me on the phone, he confided that before going, he'd dropped acid and decided to add some entertainment value after the ride and do the stunt nude.

Weeks later, with the archive footage now on hand, I viewed the VHS tape in my office as I watched Jeffery climb out of the barrel full-Monty. I had a laugh. It was pure madness, to say the least.

From there, I started to work through the list and knock out one person after another. Reaching out to a journalist, I uncovered Anne Neville from *The Buffalo News*. She had been covering the stunts for the past twenty years and helped me to better understand the nuances of them. As a traditionalist, she pointed me to speak to the father of the sport, Dave Munday.

Picking up my phone, I reached out to Dave, and to my surprise, when I got him on the other end, he seemed pretty normal in comparison to the rest of the daredevils. Dave, a purist to the sport, was now living in Nova Scotia, Canada, and was well past the years of his mastery as a barrel maker and stuntman. The time had not taken away his love of going over, which, for Dave, was an art that required skill, engineering, and teamwork. The camaraderie of the experience was what Dave loved more than the fame and recognition, which made sense to me. The end-to-end experience for him was just about being with friends, building something together, and making memories however insane.

Dave had successfully gone over the falls twice and had attempted it a total of four times. His last successful crossing was in 1993 at the age of fifty-six. At that time, he was the first stuntman to go over twice; he was later joined in 1995 by Steve Trotter. What made Dave unique, however, was that he insisted on going over in a rudimentary, traditional steel barrel with no oxygen and little to no padding. For him, it was a practice of staying true to the sport. Anne Neville agreed and expressed it in multiple articles she'd written on Dave's experience over the years.

The final and most difficult call I had to make was to the mother of Jesse Sharp. Jesse was the lone stuntman who had fatally attempted to be the first to traverse the falls in a kayak. Confident of his success, he had collaborated with a friend who was filming from the shore and with whom he made plans to rendezvous downstream later that day. Those plans would never come to pass.

Watching the footage, you could see Jesse as a little speck of a dot working his way through the rapids, charging toward the brink of the falls. I had to reverse and review the footage several times just to get a good glimpse of his canoe that was merely a speck on the massive wall of water before he launched over the brink. There on the edge, just before dropping into the abyss, he raised the kayak paddle high above his head as if to wave goodbye.

A part of me could relate to this guy taking risks without fully being aware of the consequences. I'd also done that time and time again, having been fortunate enough to survive. Watching, I wondered if he'd practiced that stunt on smaller falls, done any preparation, or just decided to go big, unaware of the risk. Whatever the case, it was a clear warning that reality does not care for your desires, and God will not save you from your own stupidity. Painful as it was to hear, Jesse's body was never found.

At the time, I had personally uncovered the footage and acquired it thorough a friend of Jesse, who'd actually filmed the event. But before I turned it over to the show producer, I wanted his mother's consent. Picking up the phone, I called, and on the other end was the fragile voice of an elderly woman answering, "Hello."

I began to speak, listened to her story, and could feel the sadness and loss in her voice. Pausing and trying to find the words, I shared my condolences and asked her permission to use the

CHAPTER 7

footage. With my nerves worn, I held tightly onto the receiver, and I quietly waited for a response.

I had worked hard to be appropriate and respect the dead. It was not my place to exploit her son, but the footage was epic; it represented her son's pioneering spirit and would make the show that much better. Then, after what seemed like hours of silence, she consented. Placing the receiver down, I breathed a sigh of relief. That was not fun.

The project *Defying Niagara* aired on the *NatGeo* network in the spring of 2007. The night it was airing, the company threw a screening party at a local bar. With the smell of booze in the air, light applause, and friends from work clambering, I got to see my name roll in the credits. With a similar sentiment of being on the silver screen, I looked at the black-on-white picture in all caps, BARRY WALTON, ASSOCIATE PRODUCER. It was surreal. I was in love with it and now just needed to figure out how I would make this more than just a contract but a career.

For the next two years, I worked my tail off on contract after contract. To further solidify my commitment to the work and company, I filled my off-time helping on random shows like *The Red Triangle*, *Escape from Alcatraz*, and *Mega Structures: The Golden Gate Bridge*. At times, I was a production assistant (PA), and at other times, I was an extra running across camera for a shot. It was dirty work for low pay, but years later, I still have friends hitting me up on social media after seeing me in an episode of *Escape from Alcatraz*. It always brings a smile to my face.

Each opportunity was growing me one step closer to living the dream. While I was losing touch with any life outside of work, I was getting where I wanted in my career. In addition, the leadership in the office, primarily women, loved me. I was a fan favorite who could do no wrong; my networking and hard work was paying off. Then, I got an assignment with a promotion to field-produce a show called, *Weird, True and Freaky* (a.k.a. *WTF*) for *Animal Planet*. It was a fast production schedule, which meant we were going to do a lot in a little time. It was going to be a non-stop six-month race to get it to air that would take up every moment of my life to produce. But if I did it right, I would soon make producer.

The show was one of the first of its kind on television, not because of the content but because of the way we acquired it. For the first time, national TV was starting to source content from YouTube. The most famous examples of this type of show are productions like *Tosh.O* or *America's Funniest Home Videos*. But this was different. This was all about the animals.

To uncover the online videos, a small team of young college grads and interns spent their days combing YouTube until they found a winner. For example, videos that showed things like a deer with six legs, a snake that bit a woman on the butt while sitting on the toilet, a farm that harvested scorpion venom, dogs that walk in their sleep, bulls with the biggest horns on earth, surviving a great white shark attack, or bear attacks caught on tape were all winners. When they found a story that we liked, I was tasked to track down the owner or people in the video, travel to the location, and capture the full version of the story. It was crazy, fun, fast-paced, and I loved it.

For the next several months, I traveled more than a pilot on an airline. I could literally be in Tennessee on Monday, Florida on Tuesday, back to San Francisco on Wednesday, and jumping on a hop to Alaska for the weekend. At points, I didn't know my head from a hole in the ground.

Before going, I needed to book the interview, so daily, I called experts and everyone involved. I would speak with research scientists, professors, and professionals in the field who could talk about the science and significance of the event too. I knocked out multiple interviews at a time from all around the country, and as the show neared its final days, I reached exhaustion.

PRIDE COMETH BEFORE THE FALL

By this point in my career trajectory, I felt pretty confident, possibly overconfident. Since having gone on the sailing trip, I had

CHAPTER 7

transformed from a scared, pathetic, hollow skeleton of a self-taught documentary filmmaker into an actual, full-fledged bona fide documentary filmmaker working in national television for the Discovery Network, no less. In essence, I had gone from a zero to a hero, and I started to make the fateful mistake of believing my own press (or thinking I was as great as they say I am). In my hubris, I began to think that I was responsible for my dreams becoming a reality. I assumed that I controlled both the idea and the outcome, and there was no credit given to anything or anyone outside of myself. At the height of this arrogance, I overlooked a major blind-spot—one that would soon rear its head and bring my entire dream tumbling down.

In the second half of the show's season, I asked for help, and help came in the form of a fresh-out-of-college newbie named Joanne. While I was looking for someone to pick up the slack, what I got was someone who was a slacker. Time and time again, she was given responsibility, and time and time again, she fell short of her tasks. I did not have the patience for incompetence or the skills to properly communicate expectations to either the young associate or the leadership; thus, I decided to plow forward with hard work, ignoring the problem and adding to the stress of an already stressful show.

In the final days of the fifty-four-episode-*WTF* production, I needed to go to Los Angeles and shoot a series of recreation shots on a very tight deadline. On this particular production, I needed to hit four locations over two twelve-hour days. At each location, I needed talent to be ready and dressed. I needed props in place. I needed the crew to be there on time. And I needed it all to happen on schedule, or we'd run over time and out of budget. It was the most stressful week of the show.

To prepare, I decided to send Joanne, my associate producer, down to LA ahead of me. Her job was to organize props, scout locations, and prep for our arrival. While I had my concerns, this was a fairly easy task, and she had ample time to execute. Three days later, when I had arrived on the day before production, almost nothing had been completed. Infuriated and exhausted from having just wrapped up another series of road trips, I went to work to make up the difference. After a full day of rushing around LA and getting props and locations finalized, I finally went to bed at 2:00 a.m. exhausted.

The next morning, things got worse. The clock was ticking, the day was full, and everyone needed to be on time. Walking out of my room, I arrived five minutes early to meet up with the team in the van. The cameraman was there, the driver was there, even the PA was there, but there was no Joanne.

My fuse, which was very short due to the stress and exhaustion, was running out. The call time was to show up to the van at 7:00 a.m. But as 7:00 a.m. became 7:05 a.m. and 7:05 a.m. became 7:10 a.m., there was still no Joanne. By 7:30 a.m., I could wait no longer. Tense and angry, I leaped out of the van and headed straight for her room. There in the middle of the hallway of a Los Angeles hotel, I lost my shit.

Caught somewhere between concern and anger, I was completely perplexed at how this person could be this incompetent and not make it out on time. All I had known to get to this place in life was work, work, and more work. For someone to be so laissez-faire about this job that I'd clawed my way up to get triggered my anger.

Pacing in the hall, I could feel a sweat bead had formed above my brow, and my right eye was twitching. I decided one more time to try and wake her from the room. This time, I pounded hard and could feel the hard wood against my knuckles. Yelling loud enough to disturb the peace, I called her name. Finally, in the distance, I heard a faint voice deep within the room.

Casually, as if nothing had happened, she said, "I am still getting ready."

What fucking planet am I living on that this girl is still getting ready while we're thirty minutes late on the busiest, most expensive days of the shoot? I thought.

Exploding right there and then at the top of my lungs, I said, "You need to get with the f***ing program. You're f***ing messing up this whole production. Get your f***ing clothes on and get your own taxi to the shoot. We're leaving without you; find a f***ing ride."

CHAPTER 7

Turning to leave, I suddenly became acutely aware of my rage. Like my own version of Christian Bale, who famously lost his temper on the set of *Terminator,* I was out of control, and I knew it. Embarrassed by my actions, I jumped in the van with the rest of the crew and said, "Let's go!" And we were off.

After completing the shoot and having returned to the office in Emeryville, I couldn't help but now notice a profound silence in the air. People who were normally happy to see me were distancing themselves. Meetings with me were short and less engaging. I was no longer the popular kid, and people were no longer turning to me for leadership. Instead, they were isolating me. In an office dominated by females, I had crossed a line and broken an unwritten rule. While I wanted to argue my case, I knew that the time to do that had passed long before LA, and now I had no case to make. Alone and being pushed out of the company, there was no way to right the ship or correct course.

When the show series came to an end a few weeks later and production wrapped, I was told that my contract with MHP would not be renewed. In the most passive and non-confrontational way possible, I was terminated—no questions asked and no one looking to resolve what had happened.

In a matter of weeks, I'd gone from working my dream job to being out of work. Now alone on the couch in my apartment, I was faced with the new reality that not only did I not have a job but I also did not have someone I loved and trusted to support me in my time of need. Different from my failures of the past, I now had come to terms with the fact that there was no one to blame but myself.

Depressed, lost, and feeling like a failure, I started to look for a place to escape. Reaching out to family, I connected with my cousin Kim, who was living off the west coast of Costa Rica. Pregnant with her first child, she would soon be leaving for the capital city San Jose to give birth—at a more modern hospital—and needed someone to house-sit. Quickly, I volunteered; it was the perfect place to go regroup.

Departing from San Francisco, the career, and the loneliness, I was now going in search of answers to where I'd gone wrong. It was a search that would lead me to places I'd never gone before and send me on a journey for love that I could never have imagined in my wildest dreams.

The hike into Big Flat on the Lost Coast of California,
my last great wave before moving to Italy

The beach Playa Aviana and the first day when I met Stefania. She's pointing at the rip down my shorts from the hollow wave that ripped them (2009).

*"I walked down a road and fell into a hole.
I walked down a road, saw the hole and fell in, again.
I walked down a road, saw the hole, tried to walk around the hole, and fell in a third time.
Today, I took a different road."*

—Portia Nelson,
There's a Hole in My Sidewalk

CHAPTER 8

THE ROAD LESS TRAVELED (2009)

MY AUNT RUTH, LIFE, AND DEATH

Throughout most of my childhood, I felt alone. It was not a feeling I could put my finger on, but there was a general disconnect with everyone I knew. Whether it was at home, at church, or at school, by and large, I had a difficult time connecting with anyone. However, there was one person I could connect with and see eye-to-eye with, and that person was Aunt Ruth.

A younger sister to my dad, Ruth was a wide-eyed and wild woman who always made me laugh. At family engagements and holiday get-togethers, she was always the fun of the party. She was a rebel with jet-black hair and a lean figure who could not be contained and always brought fashion and fun to every family event. As I grew older and more independent, I found a special place with Aunt Ruth. During the tumultuous teen years of my youth, I could call her up or stop by her house for a visit, and she was always available to listen. During our talks, she seemed to understand my hardship and always had good advice to offer.

As I finished college and moved out to LA, we'd lost some contact. She had moved to Mexico to live with her husband, and I was out chasing fame and fortune. Even with the distance, I knew I could reach out anytime, and she'd be there. Then, early morning on March 15, 2000, I got a call from my mom that forever changed that. Listening to her talk, I could tell by her voice that she was disturbed. Then, she shared that Aunt Ruth had died in a car accident. Instantly, I froze, unable to process what she'd just shared. Needing to respond, I said, "I'm going to take a shower," then hung up.

With the water running, I could feel the drips of warm water pouring over my face. Then, like a shock of lightning, it hit me, and buckling at the knees into the tub, I broke into tears. For the first time in my life, I was losing someone I loved to death and realized she would not be coming back. Sitting at the base of the shower, drenched and inconsolable, Kim, Ruthie's only child, came to mind, and I knew she was going to need help.

Thinking back as kids, my first memories of Kim were hearing her shout out cuss words like "shit" and "damn" in our house during sleepovers with my sister. For me, growing up in a very conservative Christian home, there was seldom a naughty word like this spoken without immediate recourse, so I remember this being both hilarious and shocking, and laughing out loud, I loved her all the more for it.

Over the years, Aunt Ruth would take us on trips as kids. Stopping by, she would pick us up in her white Lincoln Continental and breeze away for an adventure with Kim in tow. These family adventures were to give Kimmy some time with cousins and give my mom and dad some time away from the kids, and I cherished them then and now. Years later, after she'd bought a house in

CHAPTER 8

Mexico, I'd skipped a family Christmas in Michigan to spend the holiday in the warm sunny south with Aunt Ruth. Taking small adventures around town in golf carts, on bicycles, or snorkeling in the ocean, she'd pull out her favorite line, "Are we having fun yet?" And the answer was always a unanimous, "Yes!" Now, however, we were not having fun, and I needed to find Kim, who was in college in Colorado, and go south to Mexico to figure out what happened.

Buying a ticket, I met Kim in Tucson, Arizona. There at the airport, I remember seeing this tough young girl break down in tears as we hugged before hitting the road. From here, we needed to get a rental car and drive south over the border to their house in Guaymas, Mexico, on the west coast of the Sea of Cortez. Realizing that we were on the same road where Aunt Ruth's accident had happened a few days earlier, we stopped. Here in the middle of the desert, between two twisting curves, we found the tire tracks and got out.

Still numb to the reality of what had happened, we were observing the past events and making sense of it all. There on the road were dark skids where the tires had crossed from pavement onto gravel. There was a gash the size of a van's roof in the dirt where they'd had rolled over on one side. Broken pieces of glass with shards of plastic remained among the rocks, and I began to have flashbacks of my own accident. Having been in a terrible accident, I understood how violent the experience could be and how easy it could be to die. Aunt Ruth had been riding in the back seat and was thrown from the van. Landing hard on the pavement, she suffered a severe head injury. Her husband Don, was riding shotgun with his drunken friend, who was driving, and both had survived. I wondered whether they had struggled, like myself, to crawl away from the wreckage, find each other, and make sense of it all. I clearly imagined the horror of being isolated, alone, and having your wife dying in your hands.

The next few days felt surreal as we traveled from place to place and learned more about what had come to pass. The first visit was to the house where I'd come to spend Christmas only a few years prior. Arriving there, I could feel the memories flooding in. I could hear the paper of the presents being opened at Christmas. I could see us crammed into a phone booth in town to make calls to the family back in Michigan full of joy for the holidays.

Traveling to the hospital where her father was still in recovery, we walked into the room to find him lying on a hospital bed wrapped in bandages. Uncle Don was a giant of a man who'd worked extremely hard in his life building a paving company in Toledo, Ohio. Growing up poor, he had nothing to his name and grew the company into a multi-million dollar business. In all my years, I'd never seen a drop of emotion from him. He was tough as nails and always held it inside. On this day in a stale white hospital room lying in a recovery bed, he didn't seem as strong as he broke down in tears when he saw Kim. Pulling his white sanitized bed sheet up, he covered his face and was inconsolable for the duration of our visit.

On our last day in Mexico, we traveled to the morgue. On the back streets of a broken-down border town, we entered a poorly lit cinder brick building. The inside was damp and moist, and you could smell death in the place. A man ushered us toward the back and in a dark cold, fluorescent-lit room, we saw three giant freezers big enough to store multiple bodies. Stacked one above the other, he opened the door of where Aunt Ruth was and pulled her out on what looked like a giant cooking sheet. Opening the bag, I could hear the zipper as he opened it to view the contents inside. As she lay there, we had our first look at what remained of her pale, expressionless, and unanimated face.

Standing over her, tears were falling as Kim reached out and ran her fingers through her jet-black hair. You could see that one side of her head had been greatly damaged. There were dark black bruises and an area where the skull had been most likely fractured or crushed. Broken and without words, all I could do was put one arm around Kim as we mourned together. Uttering her goodbyes quietly, I could hear her whisper over and over again, "Mom, Mom, Mom . . ." It still hurts today to think about it. All we wanted was for this to go away, for her to come back, and for things to go back to normal.

CHAPTER 8

For me, this was the first time I'd looked death in the face, and I could not make sense of its finality. At that moment, I could feel that our time here on earth is limited and fragile. It became brutally clear that life was unpredictable, unfair, and unsympathetic to our self-interests.

Making note of her body postmortem—it was a corpse that no longer possessed the spirit of Ruth. I could only guess that the person, energy, and life that I'd known had transcended to someplace beyond. The question for me was, to where? Did she return to God, a figure on a throne in the sky? Was it that absolute? Or could it be more nuanced like the energy in the waves that I loved to surf, crashing to the beach and continuing forward in a new form?

As they slid her back in and we started to exit, there were no words to say. All that could be heard was silence from her absence.

COSTA RICA, KIM, AND THE LOVE OF MY LIFE

As I arrived in Costa Rica nearly five years after the death of Aunt Ruth, I carried a little of her with me. Hugging Kim, who was popping at the seams with a newborn on the way, I was looking forward to escaping my life in San Francisco for a while. After a few days in the country and some treasured time together, she and Dustin, the husband and father, left for San Jose to prep for the birth of their firstborn. Saying goodbye and planning to meet her again soon after the birth, I was alone to explore, surf, and take time to evaluate my life.

Having recruited a 4x4 vehicle from a friend of Kim and Dustin before they left, I also had wheels. Part of what I came down here to do was to finish the last section of the road trip with Dan, Dennis, and Zac that I missed all those years ago. They had ended the trip in Costa Rica and ever since that trip, I wanted to surf where they surfed and be closer to that missed opportunity. My first goal was to find a spot called "Witches Rock," a well-known gem in surfing because it had amazing waves, and it was one of the best surf locations visited in my second favorite surf film of all time, *Endless Summer II*. The first version, *Endless Summer,* caught me in my childhood and inspired the dream of surfing; the second one caught me in my LA years when I was on the boat and captured the age of surf that I was growing up in.

Thinking back to the Marina del Ray when I lived on my boat, I had a Sony TV tape-deck combo in which I could pop in my surf films and watch them at will. Over and over, I'd turn on *Endless Summer II* to watch the surfing exploits of Robert "Wingnut" Waiver and Patrick O'Connell, the two icons of the documentary as they surfed the globe. One of the most memorable spots on their trip was Witches Rock, where Robert August from the original film joined them to ride longboards. Famous for the giant rock that gave the place its name, I watched as they caught big glassy wave after big glassy wave, and it had burnt into my memory for all of time. Ever since then, I had dreamed of paddling out myself; now I was going to make that happen.

Jumping in the 4x4 truck, I departed early from the house outside of Tamarindo and headed north toward the destination. Sitting on the hot pleather seats, my butt was sweating with the window down as the humidity raced through, traversing my way around the tangled mix of dirt and paved roads until I came to the entrance of Santa Rosa National Park. Stopping, I pulled up to the booth, smiled at the government employee behind the window, and paid the entrance fee. Ahead, I could see the rolling, tropical green mountains. I knew beyond there was the treasure I had come to find.

Putting the truck in gear, I felt it jolt forward as I began the slow trek down toward the beach. Disappearing deep into the canopy, I was soon alone in the jungle. Tribes of howler monkeys calling out overhead were my only companions as I made my way slowly over trenches, around boulders, and through shallow river beds. Crawling at the speed of a snail with the tires gripping the edges of steep granite declines and stopping along the route several times to test the safety of a water crossing, I finally reached the base of the mountain and could feel the ocean air in the distance.

CHAPTER 8

After a full day of driving, I reached my destination and ran out to the beach. Filled with adrenaline, I could see at first glance waves were breaking, and no one was out. Grabbing my board, I raced out to the water. Paddling out alone was a little unnerving. While I loved the quiet and solitude, there was always an element of danger to surfing if something went wrong. I'd have only myself—not good odds. Again, shutting down my concerns, I accepted the reality and focused on the moment. Making it to the outside, I spent the remainder of the day surfing on glassy clear bottom waves before exhaustion drove me to shore.

On my exit, my hair stood on end, and my childhood fears of being chased surfaced. Impulsively, I stood up and ran up the beach. Taking the final steps in the whitewash, I leaped to land and could swear that something took a nip at my heels before disappearing back onto the surf. *What the fuck was that*, I thought.

Moments later, while wading up the murky waters of a river mouth toward the lagoon, my instincts hit again, and I got out of the water in which I was exploring. Walking alongside the sand, I noticed twenty yards farther ahead was a giant crocodile waiting for a meal. Fortunately, I had listened to that something deeper inside myself. Despite being alone, I felt like something or someone was looking out for me.

Checking off Witches Rock, I headed out and decided it was time to go deeper in-country. Looking at a map, I decided to drive the circumference of the Península de Nicoya. Popping out of the map in the shape of a foot, I could see that it was just small enough to traverse the perimeter in two days. Adding to the adventure, I noted another great spot for surf on the southern tip past Cabo Blanco called Playa Malpais and decided to make a go for it.

Turning south, the wheels began to roll along the paved highway, the hum was like a trance, and with hours of road, I began to drift into my thoughts. Inside my mind, I began to conjure up visions of childhood. I remembered the joy of escaping into the woods around my house. It was in that space that I could recreate myself and disappear from the order of society.

THE UNKNOWN ADVENTURER

At an early age, the chaos of adventure was something I loved; it reduced anxiety and calmed me inside. In contrast, the order and rules that came with institutions and corporations had done just the opposite and spiked my anxiety and fear. Now, in the calm of my escape in Costa Rica, it was clear why I had come and why I loved the road so much.

Drifting back into my thoughts, I began to think of college, where the biggest adventure in my life was finding my first love. That adventure was a painful lesson of indecision that led to failure and loss. For years after, I had been avoiding the hurt of a broken relationship and side-stepping my fears of commitment by chasing pointless short-term relationships with women and no long-term returns. I was addicted to impulsive hot affairs that lit up quickly and burnt out fast. Now into my mid-thirties, those choices had left me alone. Unaddressed, that loneliness had driven me toward a very one-dimensional life that seemed in part responsible for the failures in my career.

While en route to the southern tip of the peninsula, I had a moment of clarity. In a vision of sorts, I could see myself skipping gleefully through an open field, not caring about the wreckage I had created behind. In front of me, I saw a grove of endless trees with fruits and berries. Carelessly, I picked, grabbed, snapped, and snatched random low-hanging pieces of fruit before taking a bite and then discarded them all as waste. No sooner did I pick one than, like a child, I saw another and reached for it. In my wake were broken stems and damaged branches with no seeds planted for the future—a metaphor of gardening that represented the actions of my life.

Then, like a spirit, I began floating high above the orchard of trees that I had been lost in. From the air, I could see not far in the distance that the fertile field I had been enjoying came to an end. Beyond the orchard's edge, the land looked barren, dry, and void of life. Like the foothills of the mountain rising toward the sky, the fertile ground on which I gleefully skipped became hard, the soil turned to rock, and the rock was no longer tillable for new seeds. At that moment, it became clear that life in the thin air of old age did not lend itself to the cheap fleeting experiences

CHAPTER 8

of youth that I assumed would never end. Suddenly, I realized that I was wasting my youth and that once it was gone, my world would change. If I didn't wake up, smarten up, and get serious about love I would find myself wandering into a land less forgiving of my hedonistic behavior—one with no return.

With the rumbling honk of a passing semi, I snapped out of my daydream and tightly grabbed the wheel. My palms were sweaty, my breath was short, and I needed to get to my destination. Having driven for several hours, I arrived in Mal Pais. It was night, and I had not planned on any accommodation. Having few funds and wanting to save money, I stopped my car in a regional park just out front of a great surf spot. Then, I curled up in the front and fell asleep.

Crawling out of the cab the next morning, I felt rough and tired from a night rolling around without space and swatting at mosquitos. Walking down to the beach to see that the surf was sublime and having been spooked at Witches Rock by some mythical creature, I wasn't that motivated to paddle out solo again, and the poor surf didn't help.

Grabbing some coffee and a breakfast burrito, I decided to take a different route home. Looking at the map, I could see a broken-up series of dirt roads passing through the deep jungle along the coast on a route called Ruta Del Sol (106). It had a handful of river crossings, but with this off-road vehicle, I felt confident we could get over. Inspired by the adventure, I turned the wheel, and pushing the gas, I charged forward to discover what I would later call "deep waters" in my blog I shared with friends.

For the next twelve hours, I took the most unconventional route back north toward my origin, slowly creeping my way up the coast along rocky, muddy, and sometimes nearly absent two-lane roads. I pieced together the jungle. Getting out on occasion, I would step through the brush to peer over a cliff and look at the surf. Howler monkeys overhead continued to be companions. Seeing them, I'd stop and watch. Looking back, their dark eyes would point directly at me from overhead while the clan would swing from branch to branch. Standing there, something inside me wanted to join them, swinging together as a family.

Watching them pass into the distance, I crawled back inside and drove on. Along the route, I'd pass the occasional farmers walking the roadside and slowed down to wave from inside the cab. The moment also reminded me of Michigan and my longing for family.

Doing my best to make sense of the map, I traversed the final miles of the road. With dusk sitting in, I was tired from the bouncing and weaving. Having navigated the distance successfully at an average speed of 30 mph or less, it felt like there was only one way in and one way out during the full twelve-hour passage along the coast. Then, as I reached the north end of the peninsula, I ran into a problem.

In the past hour since peering at the surf, I had dropped down off a small coastal peak and into a valley. There at the base of the mountain, the road disappeared at the edge of a giant river into the riverbed beneath. In front of me was Rio Ora, a river that was running over the path nearly one hundred yards in width from side to side. This was easily the biggest river crossing I'd seen in my time driving. Listening to the strong flow of the crystal-clear water rolling out of the mountains, my mind was at a loss. I was at an impasse with miles and hours of road behind me and no clear alternate route to go home, and a decision had to be made.

Stepping out of the vehicle, I began wading out into the water to see the depth and find a course across. Feeling the cold water at my feet, I looked down and could see what seemed like a manmade cobblestone path just under the water. It was the width of the average car and dropped off steeply into deep water on its downriver edge. I inferred this was intentional and had been put there for cars to cross but its stability was uncertain. Walking farther ahead, I reached the midpoint. Here, the water remained just above my ankles and halfway to my knees. At this height, I believed the crossing was very doable and the current wasn't that bad. But just as I began to feel confident, I took three more steps forward. Here, the cobblestone road eroded in a stronger current, and the road began to drop off.

Surveying the road, I observed that the water from the majority of the crossing was at my knees or below but near the

CHAPTER 8

end the current increased in strength, and it was clearly deeper and more dangerous. Then, I could feel the chill come up around my waist as I reached the deepest section that extended twenty yards to the river bank. Here at its deepest point, the water had reached over three feet and was pulling strong downstream, making for a questionable exit.

Walking back to the truck, I was now at a clear turning point in the route. It was dusk, the light was dim, and the day was nearing an end. I had done my due diligence, scouted the route, analyzed my course, and reduced the guesswork. I was being rational versus impulsively "launching off the cliff," and now I decided whether to go forward or back.

Thinking about my options, I knew that going back would mean picking my way through the jungle at night and most likely sleeping out there. I had not seen an alternate route for some time, and all I knew was that I'd be driving the full length of the route back to Playa Malpais. Alternately going forward could be catastrophic. While the truck I was in was technically a 4x4 and did have a lift kit with enlarged rims and tires, it did not have an actual 4x4 gear. At some point in the vehicle's history, someone had removed the transmission to the front tires, most likely to save on gas. This actually made it a 2x4, and while I was pretty sure I would not sink the truck and lose it completely. I wasn't sure if getting stuck with water leaking into the cab could result in a far worse delay and a much greater problem. After all, I didn't own the truck.

Analyzing my choices, my pulse was slow and consistent as I continued to think. Then, after enough thought, I noticed that darkness was creeping a little closer and realized that a choice ultimately needed to be made. There and then, impulse and instinct took over, and I put it in gear, pulling forward past the river bank and into the water. Leaning out the driver's window, I looked down. The car splashed into the river bed, and water split around the tires as we crept forward. For the first half of the ride, I traversed the cobblestone, watching for its edge. Then, I reached the last portion and paused. Stepping out, I reassessed that we'd made it smoothly to this point. It was only about twenty yards to go on a deep dip in the road, and we were home.

Getting back in, I could hear the current rushing. Pulling forward, I watched as the wheels became more and more submerged in water. Inch by inch and foot by foot, I drove deeper and deeper until the water was above the tires and touching the base of the driver's door. As the weight of the vehicle lifted, I could feel it roll and slide. Keeping it in first gear, I pushed down consistently on the gas, slowly raising the RPMs and increasing speed without spinning the tires. I could feel the back shift but not slide. The water was swirling as it rose up in a boil under the truck. I could feel the river wanting to take us downstream. With urgency and calm, I held fast to the wheel, my hands clenched and my foot steady. With too much gas, we spin out, and with too little, we lose momentum. Both meant trouble.

With a rumble from the exhaust now bubbling underwater, we started to emerge from the deep onto the other bank. First, the tires began to surface, and the truck started shifting upward. There, I could feel the slip in the mud as I started to surface onto dry land. Soon, the back tires made it to the edge and began to climb out. Finally, with a slippery jerk forward, we lunged onto dry land.

Relieved, I let out a giant breath, which I seemed to have been holding forever. I could feel my pulse racing and the adrenaline pumping through my veins. Jumping out, I yelled "Woohoo!" On solid ground with the vehicle intact, I had made it. I could finally breathe and finish the route back toward Tamarindo and home.

Drenched in exhaustion and emotion, I needed food and a drink. Stopping in Tamarindo, I ordered food and a shot by myself at the bar. Turning, I could see a lovely young girl sitting next to me. She was a local with beautiful long brown hair and dark brown eyes. Excited, I started to share with her my adventure of crossing the river, and she was completely engrossed.

Thrilled to have company, I continued to have a few drinks, then got up the courage to ask if she was alone. To my surprise, she confided that she had a boyfriend. Looking around, I thought,

CHAPTER 8

Who? Never one to take no for an answer, I collected myself and, a little buzzed, expressed interest in meeting again.

Looking around as if to check if someone was watching, she gave me a phone number and said, "Meet me at Playa Negra tomorrow around noon." Then, with a seductive smile, she was gone.

After the craziness of the day and the madness of the night, I went home and to bed.

A NEW ROAD TO LOVE

The next morning, bored and alone, I decided to go find that girl. Throwing my board and my suit in the back of the truck, I headed toward the beach. While I'd come to this place to think and regroup, I was struggling with the loneliness and wanted a companion. While I knew this wasn't the type of girl I needed, I decided to ignore my better sense and go find her anyway.

Halfway to the destination, however, stopped at a four-way sign on a dirt road, I could hear crickets against the hum of the jungle. The spirit of Jiminy Cricket, the wise little friend of Pinocchio, was chirping in my ear. Now I was second-guessing my choice to go forward to Playa Negra. After all, I'd surfed that beach before and knew the wave. It was a little sketchy. Alternately, if I took a right at the stop sign, there was another beach called Playa Aviana that I'd never surfed. I was also thinking of the girl I was chasing. I'd metaphorically been down that road before too, and I knew where it would lead. That road was literally the road I'd been living for a decade, and to the right was a road to something completely different and unknown. In the clarity of the moment, I decided to change my direction. I turned on my blinker, pulled my tires to the right, and pushed the throttle onto Playa Aviana. What would happen next was nothing short of miraculous and would change my life forever.

Playa Aviana was a sandy beach break with a small, fast, and hollow wave. Weaving my way through a crowd of tourists and locals, I found a ride that I could surf well. Catching a thin lip on a duck dive, the wave smacked with power on my backside, blowing a hole in my shorts. Surfing with part of my butt hanging out, I was dropping into the four to six-foot faces and getting covered up inside a barrel time and time again. Inside that hollow space, the sea seemed to wrap you up, and time stood still while outside people watched you fly by. In there, you were in complete isolation, experiencing a reality from another planet.

Worn out, I needed a break and paddled back onto the beach and walked up to a small restaurant and bar. Grabbing some pizza and a drink, I decided to sit down and read my book beside a few people who were enjoying the day under a palm.

While eating, I couldn't help but note that the four people I'd sat next to were smoking cigarettes and speaking in a foreign language other than Spanish. They were all dressed clean, even in their bikinis, and had Gucci sunglasses and Versace sarongs that felt classic in a way that American wear did not. Observing closely, I realized that I knew these people; they were Italians, and of the four, I noticed one very cute girl sitting in the middle who I wanted to meet.

Reminiscing, I thought back to two years earlier while living in Santa Cruz and teaching. With summers off, I had picked up some extra work as an English as a Second Language teacher. Santa Cruz High School had an ongoing exchange program with students who'd come over from Italy to learn the language and spend time in America. Having known Italian culture only from TV and movies, I was excited to have the chance to work firsthand with these students.

Instantly, I fell in love. The spirit, the friendliness, and the welcoming nature of these Italian students was amazing, and I couldn't get enough. One day, a student shared a piece of his mom's homemade beef jerky from a small bag he'd transported from Italy. It smelled of delight and tasted of something otherworldly. One bite, and I loved it. In addition to the love of food, I noticed the closeness of the family between these kids was unlike what I knew of in America. There was a connection here with

CHAPTER 8

family and a love for good food that brought people together that I understood and valued deeply. It reminded me of the fish fries on my grandpa's farm and inspired more curiosity of this culture and its people.

Back on the beach in Costa Rica, I observed these folks while ants were devouring my pizza. They were very close to each other, spoke fast, and had that unique connection that I envied. Then, when three of them snuck away to the beach and left the cute girl back alone, I spotted my opportunity. Moving closer, I said, "Hi." And in one breath, I went from observer to participant.

Smiling, she said, "Ciao." Her name was Stefania, and right there, someplace between the broken English, Spanish, and Italian, a conversation broke out.

The connection was instant, and the conversation was a daylong game of words and charades, but we found our way. In one day, we spent a lifetime together. We took walks on the beach, swam in the water, and traveled through space and time in our eyes. With magic in the air, time flew by, and before we knew it, the day was over.

Like a scene from a movie, the sun began to touch the horizon, and golden light began to settle in the sky. Like the breeze passing through the palms, the day was over, and our time was coming to an end. Trying to convince her and the group to join me at my cousin's house, I chased them back to their car, but Stefania, a good Italian girl from a strong family, couldn't go. It was too soon, and the group was on their way south; my place was north.

Thinking on my feet, I grabbed a pencil, opened the book I had sat down to read, turned to an open page, and asked for her contact info. Sharing, I wrote down her email, number, and as a second contact source, her Facebook page (which was in its infancy).

Then I said, "I'll come visit soon."

She smiled, thinking I was crazy, and said, "Ciao," and, in an instant, had left as soon as she had come.

In her absence, a void rushed back in, and I realized more than ever that I was completely alone but with one difference: I now had a mission. My mission was to find the girl. And I was willing to travel the globe to do it.

On my last days in town, I visited my cousin Kim and her newborn son in San Jose. Walking into the delivery room holding her new baby boy, I could see her face glowing with joy. Next to her was her husband Dustin, standing proud and looking on. Standing there, thinking about the new romance that had just blossomed, I knew now more than ever what I wanted for myself. I wanted a family, and I had someone in mind that I could do that with.

Standing over the Roman Forum in awe at my first visit (2009)

"True love will find you in the end, you'll find out just who's your friend. But don't be sad. I know you will but don't give up until. True love will find you in the end."

—Daniel Johnston

CHAPTER 9

WHEN IN ROME (2009)

ENCHANTING LOVE IN ITALY

Imagine yourself sitting at a long wooden dining table in a kitchen that feels like something out of a scene in the film *Gladiator* in ancient Rome. In the center is a bottle of wine, a loaf of bread, cheese, and an assortment of cured meats. Ahead of you in the distance through the large open French doors is a slice of Tuscan countryside, filled with fields of golden brown wheat and blossoming yellow sunflowers. Cutting through the middle of the valley is a Eurail train racing past at 100 mph. All around the hillside are scattered plots of old brick and stucco farm homes, and on the hilltops sit ancient city walls protecting cathedrals, spires, and castle towers. The sun is pouring down, a light breeze is rolling in, and you are at the center of it all. If you have imagined this moment, then you will be able to imagine my life just six short weeks after I returned from Costa Rica.

After leaving Stefania on the beach and saying my goodbyes to Kim and Dustin, I returned to my San Francisco apartment and gathered my thoughts on all that I had experienced in the month down south. Still unemployed, I now focused intently on the beautiful Italian girl I had met only a few weeks earlier. Having searched and found her profile on Facebook, we began corresponding, and I made plans for a visit.

Needing no further convincing, I looked on the travel site Expedia.com and impulsively pulled out my credit card to book a flight to the old country. I can't arguably say that I had the funds to justify the cost. What I can say is that I needed to go and find out if I had discovered true love. I knew that if I didn't answer that question, it would linger in my head for many years to come. After a few days back on the ground, I boarded a plane and put on my iPod to listen to some of Rosetta Stone's course on speaking Italian and took off to the old country.

Driving in her Smart Car, Stefania picked me up at the airport, looking like a Roman goddess. Finally, we saw each other again in her country. Instantly, it was clear that the connection was still there. As we raced through the streets of the eternal city, I was enamored with her every spoken word in English cut with an Italian accent. After weaving through traffic and squeezing into small streets, we settled into our bed & breakfast in the middle of the city and started to plan our time together.

The Ancient city of Rome is romantic, divine, and powerful. Over 2000 years, it has developed into a winding twist of roads tangled among towering walls and aqueducts, glorious historical structures, fountains, statues, and a scatter of Obelisks taken from Ancient Egypt. Taking to foot, we'd walk upon the Pantheon with its sixteen Corinthian columns made of Egyptian granite and its oculus (central opening) at the apex of its perfect dome or the Colosseum with its seemingly endless arches, elliptical shape, and superior Roman architecture seating some 80,000 people in its height. Just next door was the Roman Forum, a center of public life in the empire with its Triumphal Arches and the Curia, where the senate convened to pass laws. Each one took my breath away and forced me to pause in my shoes

CHAPTER 9

while I processed what had been a seemingly mythical history coming to life. Over and over in my mind, I said to myself, "The Romans were actually here." Intertwined in my walk, I began biting into a slice of pizza or twisting up a mouthful of pasta. The food stimulated taste buds I didn't know that I had and forced me to question everything I'd once ingested prior to this. I had to question if what I had been eating back home was cardboard in comparison. With each bite and each step, I was discovering a part of the world I never knew existed.

It was already mid-afternoon by the time we'd stopped walking and sat down for an aperitivo. The waiter brought a sparkling drink of Campari and Prosecco with a navel orange out the top called a Spritz. Behind me, there were vines walking up the side of a crumbling stucco wall. And in the distance ahead was Saint Peter's Basilica standing high with its iconic dome designed by Michelangelo in the Renaissance period. Looking at Stefania smiling, it was noticeable that she was proud of having shown me her magnificent home. Ahead of us was a few days of adventure beyond the walls of Rome. From here, we were headed north to Tuscany to tour the land before ending our time together at her mother's house on the outskirts of Rome's countryside.

After a day spent walking through the cobblestone streets, I was standing atop the balcony and gripping the iron railing. Breathing in the air of flowers, I couldn't help but note how alive the city felt. Women were hanging clothes to dry, kids were playing in the square below, and dads were drinking wine and smoking cigarettes on the corner. On every balcony, plants were growing over, dangling down into the space below. This place possessed family and culture and was alive in a way that no American city was.

Departing for the golden hills of Tuscany, we jumped into Stefania's ride and took off. Gripping the wheel, she was at the helm, and I was her passenger. Shifting us into gear, we began cruising through the countryside in our little bubble with windows. Pressing my face hard toward the glass, I twisted my neck up and around to see small villages appear on cliffs above. These structures perched high on a rock face or wrapped atop a mountain peak, protected completely by walls and gates, seemed strangely fictional, but people were still living, worshiping, and dying there. Watching at the speed of a car cruising on the highway, I wondered what the world must have actually been like for people to build and live in such closed-off spaces. I thought about the human experience and the factors that drive people to close themselves in and protect their resources from the dangers that lurk. Observing the scale of the walls, it felt like the dangers must have been very severe, but did they warrant such dramatically, enthralling, and challenging places to live?

While I knew in theory that it was the Dark Ages that drove this, I also knew that at some point, the Dark Ages came to an end. At that point, people could leave, but many did not. It was here at this crux between staying and going that I began to ask, why do people live in the confines of walls of all forms (i.e., family, religion, relationship, job, and corporations) long after the dangers that drove them there are gone?

It made me think of my own life choices and the walls I'd gone beyond. Doing things like moving to California, chasing my Hollywood dream, driving and sailing for months on end, or now traveling to Italy in search of love were all things far beyond the walls I'd known. Doing so was scary, hard, and sometimes painfully wrong. But as a course of action, I was free to live my life. As we moved farther down the road into new spaces unknown, I wondered if the experiences of the first villagers to leave these walls were like mine. Did they leave people they loved behind feel lonely and isolated from the culture and desperately long for people to understand their reasons for going?

Halfway to our destination, Stefania was focused on the drive. Turning, I observed her strong, toned bronze arms with a golden shackle hanging from one wrist. Happy as a peach and excited to show me more of the country, we were now free of the walls of Rome, free from any walls in theory, and on our way to a completely new land.

Traversing our way through Florence, Siena, and Proto, we entered under towering gates into beautiful cathedral squares

CHAPTER 9

and over some of the most romantic of bridges in the world. We'd wander into the quietest of alleys to find ourselves in the smallest of restaurants that had been built right into the walls of the city. There tucked away in a corner, you could smell the most amazing foods being prepared—classics like pasta Amatriciana or pasta Carbonara were on my simple mind. But instead of ordering myself, I turned the ordering over to Stefania, who would order Zuppa di Mare, Tortellini Panna e Cotto, Fusilli con Pesto alla Salsilliana, or an entire list of foods I would have never even known existed. Whether it was prosciutto crudo, fried Roman artichoke, fagioli con trippa, or any other long list of foods, I was always satiated with every bite, and my mouth still waters today just thinking about it.

As we sat there eating in those first days, we were still stumbling over words and phrases between the two languages. But we were working hard and getting better. I would speak slowly and clearly in English, and she would offer up new words in Italian that I hadn't learned before. Yet, beyond the language, there was something bigger being communicated—love. While the barriers of culture, location, and language were big in comparison to most, the love that was growing for each other was far bigger, and it had me thinking of something I'd never imagined before, moving to Italy.

At night, while she was sleeping, I would lie awake, working on how to put together the pieces of a move to Italy. I was currently out of work, post my falling out, and living on savings, so, financially, I had no business considering this. In total back in America, my physical possessions boiled down to only a few things. Post all my travels and moves in the past years, I had grown very light on possession and had nothing other than a pickup truck, my Sony FX1 camera, my Apple computer for editing my documentaries, and some clothes in my closet. This meant that physically moving would not be that challenging; however, there was more to this than the physical. While I was never big on possessions, I also had not built any wealth for the future, and I wasn't getting any younger. And it was here that the conflict was born. I knew that without moving to Italy, this trip and romance were purely a fantasy that would soon come to an end. However, if I did move here I would be starting over, leaving behind everything I had in my finances and friends.

Between the two choices, I could also hear a little voice inside me saying, "Keep going, never stop dreaming, carpe diem (seize the day), and don't fear the future, for the future is yours to take." Working to balance my romantic, spiritual, and creative side, I began to ask myself more rational questions like, "When are you going to grow up, get serious about life, buy a home, start a family, and work a solid job?" All questions the adult in me felt I needed to start to answer. The reality was that the move was not rational. It was going to be a hard road, and there were a lot of unknowns.

With only a few nights left on the trip, we returned to our house in Tuscany. There, I sat down to eat at the table overlooking the golden fields of sunflowers, plots of farmhouses, Eurail trains, and castles in the countryside. Pausing for a second with a glass of wine, I felt my romantic side winning the argument and decided to give it a voice.

Standing straight against the beam of sunlight peering in the window from above, Stefania was cutting tomatoes and preparing pasta when I asked, "What do you think if I move here?"

Prior to this moment, everything had been fun and light. The truth was it was a game of playing house at best. But this question was a turning point, and the tone and mood changed. Now the full weight of reality was underfoot, and we would soon determine if this little seed of love was going to pop above the surface and live or die before reaching the light.

Thinking back on the road that it took to get here, I could see every big risk that I had ever taken and every lesson that I had ever learned. Over the past few years, I became a Ph. D. in travel, adventure, and the struggles of making a dream into reality. Now, it had all come together for this moment—a moment that would define my life forever. Whether I could straddle the gap of culture and time, survive the transition and struggles, and return victorious was all in question. But I had taken the risk, opened my heart, and put my cards on the table.

CHAPTER 9

Pausing to see if I was serious, Stefania responded as lightly and unstressed as she had at every turn in this journey. "Yes," she exclaimed. "Great! You can get an apartment in Rome; I can help."

Instantly, I rejected the idea and clarified, "I am not coming here to be a tourist," I said. "I am coming here to be with you."

Then, like unlocking a safe full of gold, Stefania's heart opened up. There and then, she realized the depth of the scenario and the seriousness of my decision.

In my last days in Italy, we traveled back to Rome to visit her mom and sister. There, over some amazing food and wine, we shared our plans with family who were elated and embraced my return. This was somewhat new for me—a family that welcomed my bold risk with a brave embrace and further inspired the cause. That night, over a glass of her late father Pasquino's best whiskey, we celebrated. The next day, I was on a plane back home to America, and now the clock was ticking.

Photo taken on our trip to Paris, where I ultimately proposed (2010).

*"It's not where you begin in life;
it's where you end up."*

—Zig Ziglar (paraphrased)

CHAPTER 10

LOVE, MARRIAGE, AND LIVING THE DREAM (2009)

GOODBYE, CALIFORNIA; HELLO, ITALY

A few long flights and a day or so later, I was back in San Francisco recovering from all my travels. Alone in the quiet of my apartment, my life in California, which I had been away from for some six weeks, was now coming back to me. Photos full of friends, adventures, road trips, and surfing hung on the walls. Around my apartment were surfboards, skis, snow shoes, hiking boots, and all the stuff from all the outdoor adventures that I learned and loved so much from living here.

Relaxing on the side of the bed, I looked outside to see the Pacific fog had moved in over the Inner Sunset, blocking my view from my panoramic bedroom window. Squeezing my pillow between my arms, all the memories from my time in California came rushing back and grabbed ahold of me, begging me to stay. Now I realized all that I would be leaving behind. This was a big leap like no other that I had taken before in my life. I could feel the fear of what I'd committed to do sinking in, and I began to hesitate.

Inside of myself, I knew that the emotion of fear is a powerful defense mechanism and could easily overwhelm my decision-making. It was said best in the movie *Point Break,* when Bodhi (played by Patrick Swayze) said, *"Fear causes hesitation, and hesitation will cause your worst fears to come true."* I could relate.

Designed to protect, fear grabs hold of you and sends you into either fight or flight mode. It sees the world as friend or foe. It reminds you of your mistakes and weaknesses and drives your limbic system hard toward pausing to make the safest choice. That feeling can be paralyzing, and it had the ability to end everything that I was working toward. However, I had a plan in place to overcome fear. In my school work during college, I remembered a lesson on changing behavior. At the core of that lesson, it taught that habits of behavior cannot be removed; they can only be replaced. For me, in this scenario, I decided to replace fear with clarity. After all, I'd had a moment of clarity on the beach in Costa Rica and was very clear on what I wanted to do when I was in Italy. Now, I just needed to apply that clarity to what I wanted for the future and keep going forward.

Over the next seven days, I called my landlord and ended my lease, informed my roommates, visited friends, then packed and loaded all my possessions into the back of my truck, and pointed the vehicle east toward home and hit the road.

On the miles across the country, I started to think of Italy. Gripping the steering wheel and gazing into the open deserts of Nevada, memories of Stefania and the time we'd spent in the old country rolled through my mind. I imagined the dream of our new life together and everything we'd do—the dinners around the table, the time touring the city, and her cascading through a field of sunflowers with the golden light of dawn all ran through my imagination. A complete fabrication of reality, but it was the escape I needed to sustain my imagination and drive hard onto Montana, Wyoming, and across the country at the pace of a jackrabbit.

CHAPTER 10

In Montana, I slept under the truck in a sleeping bag on BLM (Bureau of Land Management) land to save on costs. I decided to hold my favorite ice ax next to me for protection. In my mind, I wanted to be prepared for a stranger, a wolf, or a stray dog that might show up. Fortunately, none did, but as I woke at the break of dawn and scrambled to keep moving, I left my ice ax on the ground as I pulled away. Some two hours farther down the road, I realized it was gone and squeezed the wheel hard. With white knuckles, I yelled out, "Damn it," hitting the steering wheel dramatically as if I'd lost someone I loved. I knew it was only an ice axe, but it felt like a physical representation of all I'd left, and it hurt to let it go.

When I arrived early in the morning at my parents' home, Mom was still having her coffee. When they saw me, they came out of the house to greet me and gave me a big hug. Having seen me now through multiple wild and impulsive moves, my parents were just happy to spend some time with me. Being older, they'd developed a thicker skin about my comings and goings. I guess they'd accepted the reality of who I was and, with that, stayed present and supported me throughout the process.

While there, I went to my Dad's Windows computer to log into Craigslist. I punched in the details of my little white Toyota Tacoma pickup truck with the sleeper camper on the back. It had been my biggest treasure while out west—a loyal companion on many an adventure. I'd driven it around from surf spot to mountaintop. To let go was hard, but again, I had no choice. I needed the money. A lovely granola-eating couple from northern Michigan came down to purchase it. I knew it was in good hands, counted the $8,000 in cash, and then bid the truck farewell.

With the money, I bought my ticket, packed the last of my belongings in my parent's basement, and boarded a flight back to Rome. Sitting snug in economy class with my knees nearly flush against the row ahead, I finally could pick up my head and breathe. Letting out a sigh, I was proud to know that fear had not won over me. I did not flounder. Pressing deeper into my seat, I watch flight attendants work the aisles. Behind me was the solo life that I'd built on the West Coast. Ahead of me was something completely new. Halfway across the Atlantic, I put on my Sony headphones, scrolled to play on the iPod, and started listening to Rosetta Stone's lessons on speaking Italian. I was on course to an adventure that would change me forever.

BEING AN IMMIGRANT IN A FOREIGN LAND

Being in Italy as a tourist is much different than being there as an immigrant. In going there to live, you are no longer there for escape, pleasure, and the romantic notions that you normally build on a fun visit. Instead, you are faced with the day-to-day realities of work, routine, and survival, just like back home.

In the transition to our life in Italy, I definitely had a small, extended period of euphoria. However, this period was followed by serious culture shock and realizing how hard it is to function in a foreign land.

The biggest hurdle to getting started was the language, which was really, really hard to learn. Even buying a drink, having a slice of pizza, or just saying "Ciao" on the street required some knowledge of how to speak. Suddenly, like a slap in the face, I realized I was at a great disadvantage without this mastery. The mobility I once had in society was no more, and I desperately needed the language to function. Fortunately, for me, I had Stefania, but she could not always be there, and when she was, sometimes she needed a break. After all, I was completely dependent on her for nearly everything in those first days. Daily, my brain was exhausted by a barrage of new words, the tongue-twisting enunciation, and work to form them into sentences that made sense, but survival, the great motivator, drove me forward.

Possibly more important than language was shelter. I quickly discovered that life in Rome was not cheap. To my good fortune, in preparation for my arrival in Italy, Stefania's family had gotten us a small 700-square-foot apartment just down the street from her mom. That small place with a bedroom, living room, bathroom, kitchen, and glorified closet-office studio cost us just over 1,000 euros per month. Mind you, the average Italian brings

CHAPTER 10

home about 2,000 euros per month in income just to live so that rent was at least half of our combined income. Still, for a small space and a big price, we had a home.

The last big piece needed to survive in Italy was work. While I had assumed that my film, television, and media career would transfer over, I had assumed wrong. The film and media industry in Italy is far smaller than in America. And the last thing they need or want is a non-native English-speaking foreigner coming over and stealing a job, especially in an industry and market as competitive as mine. In time, I would work on films as a volunteer. For now, to survive, I had to accept a job as an English as a second language (ESL) teacher at the British Institute in downtown Rome. In a few short weeks, I had resorted to what I knew best, speaking English, and was now established enough to survive.

In hindsight, like on the sailing trip and so many trips before, I had moved to Italy with an idea of what life would be like that did not match up at all with the reality of what life was. In my defense, there were no manuals for the move to Italy—no YouTube tutorials. At best, you could ask for the advice of friends, and I did, but even that advice was patchy.

What I got was, "Bring an electric converter for your power, your camera for your hobbies, and remember, it gets surprisingly cold in those old brick-and-mortar buildings in the winter."

That was as good as it got. The truth was that the transition to life in Italy was hard, and I was soft.

Like so many in my generation, I was spoiled by the life that people around the world wanted so desperately to have. Throughout my life, I had long overlooked the words of people greater than myself who would preach how we did not appreciate the quality of life in America versus the rest of the world—how Americans were too demanding, ungrateful, and going through life with too high unrealistic expectations. I, however, was an immigrant in a foreign land working to survive and beginning to find out firsthand what life outside America was like.

Living in Italy, I began to discover the dysfunction of things like the high cost of living, the broken systems of government, the painfully long lines everywhere, and the failure of immigration to keep my paperwork from getting lost. However, the worst day-to-day thing that impacted me directly was surviving on the public transportation system that was needed to get to work. I remember being stuffed tighter than sardines on a bus holding in my hand my little flip phone tight to my ear and yelling over the crowd of people, trying to ask Stefania for help. I was stuck, unable to get to work, the train was shut down for a strike, and for everyone around me, this was just another day in Italy.

Prior to moving, I'd been spoiled by the fact that living in America, if you can speak English and want to work, you can make money. But I was learning the hard way that since I was no longer a native-speaking citizen, the only job available for me was the one paying ten euros an hour teaching English in the middle of Rome. The only problem was the commute. To get there daily, I needed to take a bus for thirty minutes to a train, a train for thirty minutes to a transfer, and a transfer for ten minutes to a stop, and then go on foot for fifteen minutes to reach the school. It was a one-hour and thirty-minute haul one way, and that was on a "good day."

On a bad day, there would be an aforementioned transportation strike. That was when the transportation union of subway train drivers would randomly announce that they were not coming to work. This announcement usually happened without enough time for anyone commuting to work to react. It wasn't until I arrived at the train station that I discovered a strike was happening. The first indication would be an unusually large number of people waiting to get on the bus versus traditionally everyone getting off. When I first experienced this, I exited my bus and walked down to the station to discover the second and more obvious indication of a strike. The train station was closed, and a big steel cage-like door spread across the entrance, blocking your passage. On that day, confused, I joined the thousands of other Italians working to enter a bus downtown, then pulled my flip phone out and called Stefania.

If the strike wasn't disruptive enough to the flow, the process of riding a bus was beyond anything I'd ever experienced.

CHAPTER 10

Standing in line with hundreds of people, I looked around at all the faces. Like me, immigrants from Albania, Liberia, Morocco, Egypt, the Middle East, and Europe were all pushing into buses, maxing out the load limits. I could feel the man behind me shift from side to side or the woman in front of me lean backward into my hips as the bus hit its brakes. Bumping hands with strangers, I reached around to check my pockets for valuables, then held them close to my chest. Nudging the person behind, I pulled my backpack around to the front. I didn't want to ride in a compressor for thirty minutes and also get robbed. While entering and exiting, people would have to collectively lean inward, with the last members swinging forward on handles while the auto door attempted several times to close completely. Exiting, I felt like carbonated water popping out of a soda bottle. Bubbling up, I no longer could differentiate between people; it was just a fluid gush out the door. No easier time to pick a pocket than now, I thought, stepping onto solid ground. Outside, I finally could breathe a sigh of relief and decompress.

Time and time again, these experiences and others reminded me that I was an immigrant in a foreign land. Surrounded by people from other nations, I wanted to think of myself as special, but I was not. I could live and die out here on the streets as easy as anyone, and there were no special services just because I was an American. In those early days, I dreamed of my life in California, where I had had it so easy. At times, I wanted to go back, but now I had bills to pay and a life with someone else who I was working to build. So, day after day, I woke up and went to work.

It wasn't all bad, however. I learned incredibly valuable things about life, and the trade-off for hardships shared among people seemed to be a kind, understanding, and forgiving Italian culture. I mean, there was something very bonding about the struggle with Italians, and the struggle felt universal. People looked out for each other in a way I don't remember experiencing in America. The words "non preoccupato" ("don't worry") were like a motto that while, at first, I didn't understand, in time, they began to make sense and reduce the anxiety that I'd always known back home.

Walking into the house at the end of a long day, I could smell the fresh pasta sauce. The aroma of tomatoes boiling with garlic, basil, and salt filled the room. The table was set with a cloth, and on top was bread, cheese, and some slices of salami. Taking a bite of pane, I could hear the crust crackle in my teeth, and someplace between the artisan flavor and the light texture, my struggles seemed to disappear. While the language barrier was hard and the land foreign there was something very human about this experience at home. We did not have a battle for sexual identity or fight over our individual roles like the "liberated women" of America. There was something far more traditional and balanced here. The distinction in roles inspired me. I wanted to do more to contribute to the family, be an honorable man, and help build the home.

After having been in the country for four months and working extremely hard to make ends meet, I specifically remember experiencing a feeling, something I'd never felt before. Exhausted from the commute, I laid down on the couch and looked up at the ceiling. At that moment, I felt a sense that I was home and had family. With that feeling, I was able to put down the incessant burden of anxiety that I had been carrying in solitude for so long in America. Here, I felt taken care of by the culture and looked after by the people who surrounded and loved me. The stress that had once bound me unraveled inside like a ball of yarn. I was truly falling in love with this country, these people, and their way of life, and nothing was going to take that away.

KEEPING THE FILMMAKING DREAM ALIVE

Outside of our relationship, work, and life in Italy, I still wanted to create documentary films and, in general, do filmmaking. Before leaving to go to Italy, I had packed my Sony FX1 camera with which I had captured everything I shot on the sailing trip. I had also spent what little money I had on a new laptop computer

CHAPTER 10

with the software, Final Cut Pro, with which to edit. When I wasn't teaching, I was also volunteering on sets for films, one of which was *The Violin Ghetto,* a story about the struggles of violin makers in Cremona, Italy, known as the home of the violin.

Joining with two Italian brothers, Giorgio and Marco Priori, the owners of NextSun Productions, I started working and traveling north to the town of Cremona, Italy, "the home of the violin." There, a group of students were living in a broken-down block of apartments, struggling their way through violin-making school. Climbing into the tight quarters of our subject, Klaus's workshop, I filmed the UV-B and UV-C light rays drying the varnish of a new cello he was making. The blue light rang into my lens, and Klaus quickly shut the door after only a few seconds of footage to shield me from too much exposure. Then, breaking for lunch, Giorgio let out a big chuckle of enjoyment as he picked up a bottle of wine and poured us all a glass. This was more a theme of the project than the project itself. After all, we were in Italy.

This work, more often than not, was for love and experience without the pay that I desperately needed. Needing more money to survive, I had an idea. During my time in San Francisco, I'd followed and completed filming a documentary on the UltraRunner Mark's running of the *HURT* 100 race in Hawaii. Having put together a small edit station in the studio room in Italy, I started to digitize the tapes and decided to complete the edit. The hope was to sell the DVDs for money. The idea was inspired by Marco and Giorgio, both of whom were amateurs like myself and had recently completed a film on modern-day gold diggers shot in the Yukon called, *Yukon, the Last Rush.* Sitting in a room on a cold folding chair with a group of thirty other people, I watched the two brothers screen the film on a drop cloth with a projector. It was after listening to them speak at the Q&A that I realized for the first time I could do the same. While my sailing documentary was still unfinished, I decided it was time to take a risk and put my work out there. Returning to the cold tile floor of my small studio office, I went to work on creating my first documentary called *Profiling HURT.*

THE UNKNOWN ADVENTURER

Punching on keys and cutting with clicks, I pieced together the story, which centered around runner Mark Gilligan, his extreme obsession with running, and his quirky way of training. Back during filming, I had been commited to capturing the full experience, I decided to run-walk a full twenty-five-mile lap of the four-loop course. Moving through the passing showers, I could feel the leaves of the Giant Taro plant rub against my body as I worked my way through the jungle for eight hours. Sore to the bone, I felt the exhaustion and struggle as each runner traversed up and down the mountain course, slipping and stumbling across the maze of roots and rocks for the thirty-six hours of the race. Working to capture the first person "point of view" (POV), I rolled the camera as runners brushed past me or led in front of me on their course, working to make the cut-off time. Showered with golden light, I was overcome watching the sun setting, rising, and then setting again over the course of filming the race. Over and over, the athlete's summit peaks at 2,400 feet, drops to an aid station at the base of the mountain, then summit again, three times per lap—a total elevation gain of 29,000 feet—the equivalent elevation of Mount Everest over the course of the run.

Looking over the footage, I was reminded of all the hours I spent working to capture the experience, but through the pain of blisters, bruises, cuts, and cramps, I'd captured it. For the runners, I discovered that the experience was a microcosm of life—a snapshot of what it took to make it to the finish line of time. As I edited together the story of them crossing the finish line, you could see the relief in their eyes from the exhaustion and the enlightenment from what they had completed. It was inspiring and crazy as hell.

Reflecting on the experience, I'd come as close as I could as a journalist to feel what they felt, and I was happy to see it complete. I avoided writing voiceovers (VO) to tell the story and edited the project in "verite"—as it unfolded—allowing the minds of the viewers to piece together what wasn't told in visuals. My goal was to capture the feeling, struggle, pain, and chaos of the course.

CHAPTER 10

Proud of my small accomplishment, I had finally done it, and after three months, it was finished and ready to show to people. With the goal to make some money, I had 500 DVDs made and shipped to my parents' house in Michigan. Commissioning my mom as my first employee, she agreed to help ship them out as orders came in. In the end, I'd sold nearly every DVD, making a profit of around ten dollars per sale, a total of $5000, much-needed cash.

In the process of making them, I shipped one copy to the *Sedona Film Festival* in Arizona. It was a long shot, but a friend who helped with the event recommended that I enter it. When I was accepted, and I received a Laurel from the festival, I felt a sense of significance and validation. It came with a sense of achievement that I hadn't gotten before, and I wanted to go further and do more.

EUROPE, THE SCHENGEN TREATY, AND THE WORK VISA

On a Skype call with my family over a distance of thousands of miles, Stefania and I squeezed into the frame to say hello. It was Christmas, and nearly six months had passed since I'd moved to Italy. Seeing the smiles on my parents' faces, I pressed play and watched as a holiday video I'd made of my time in Rome bounced off satellites and across borders to arrive at my parents' house. It was a special moment, then we wished a Merry Christmas and said "ciao" before hanging up.

Running into my first major hurdle since moving, I discovered in the New Year that my tourist visa would expire. Being on a traveling visa, customs will allow you a six-month grace period in which you can work. After those six months, you need to have either a Permesso di soggiorno (resident permit) —which you could only get through corporate sponsorship, marriage, or a stamp on the passport from traveling through customs from abroad. My employer at the British Institute, where I taught English, pulled me into his office. Looking at me sternly in the eye, he wanted to emphasize the seriousness of the situation if I didn't renew my visa, I would be fired. From this point, I had a month to solve the problem, so I went to work.

Since I was forced to leave the country, I decided what better place to go to than Paris, France. Not only would it be an amazing place to visit, but it was also a cheap flight as Ryan Air was offering deals for as low as eighty-seven euros from Rome. For this price, I could fly both Stefania and myself, spend a few nights there, see the Eiffel Tower, and return through customs with my stamp.

Traveling to the most romantic city in the world would be rare, unique, and a one-time-event experience. Then and there, I had an idea. Stefania and I had been together now for six months, and things had been going amazingly well. While the move to Italy had been somewhat impulsive, my instincts had been right. Now, I wanted to go one step further; I wanted to propose. And where better than in Paris on top of the Eiffel Tower.

Having made up my mind, I decided I'd first ask Stefania's mom for permission before actually proposing to Stefania in Paris. Her father had passed away some time ago, so her mom was the only parent left to ask. We were not kids anymore, so it was somewhat of a formality, but to me, it made sense to show my respect. To do this, I planned to ask her mom during one of our weekly lunches, which were a highlight since moving to Italy.

For a few days a week, I'd not have to teach and would remain home from work. On those days, Anna, Stefania's mom, would invite me over to eat. At first, it seemed like she was being kind, but with time, I realized that not only did she enjoy my company and loved serving food, but it also represented a duty to family that I had not experienced before. So, once a week, I would walk the mile or so from our place down the hill to her mom's to eat.

Anna was a Sicilian woman, and if you know anything about Sicily, you should know their cooking is the best in the world. My mouth would salivate as I walked in the door to the aroma of a pot sizzling in the oven. The flavors of that dish would pass through the house in waves, and the place was like something from the best restaurants in New York City. Sitting down, I

CHAPTER 10

squeezed myself into the corner of her little kitchen on a small thatched wooden chair, pinned in the corner against a white tile wall. I would lean forward over a plate, and my eyes would fill up with the idea of what was to come. Things like fiori di zucca, polenta con spuntature di maiale, tortellini in brood, and much, much more. All family recipes are made by memory.

On occasion, her companion, Angelo (aka il Padrino), would join. On those days, he'd bring his homemade sausages, cheese, and personal wine as an aperitivo before we ate the main dish. Lunch could go on for nearly two hours; I was in heaven. It was a pleasure unique to Italy.

On this day, with a proposal of marriage on my mind, Angelo would not be there. Walking to Stefania's mom's place, I could feel the pavement beneath me as I was stirring over the idea inside. Mentally, I had decided I would wait till after I ate. Then, when the moment was right, I'd ask for her daughter's hand, which I thought was a very romantic gesture. Flipping through the pages of a small pocket-sized Italian dictionary, I worked to memorize what I wanted to say before I arrived.

When words of a new language come out of your mouth, you cannot feel the context of them, so at times, it feels like you are babbling. You know you're saying words, and you see people listening, but the context and history stapled to your mother tongue is not there. Instead, you fumble around like a baby, and much like that baby, more often than not, I talked like a four-year-old child. The process left me feeling incompetent, like returning to elementary school all over again. But unlike the past, I was determined not to fail.

Engulfed by the flavor of the pasta fettuccine con salsiccia fungi piselli (pasta with sausage mushrooms and peas) with some shredded Parmigiano Reggiano, I ate up. Then, I grabbed hold of the tiniest expresso glass for one final bump before deciding it was time to ask. Pushing back from the table, I rubbed my belly, which had grown significantly since I'd moved there. Across the table, I looked at Anna peeling an orange. Handing me a piece to eat, I took a bite and tasted the tangy fruit as I quickly pulled out my Italian dictionary one last time to refresh my memory on what to say. I knew in this case that I couldn't elaborate or share much outside of the script I'd memorized. Once I started talking, I couldn't share with her the cultural significance of asking for her daughter's hand. I couldn't make adjustments if things didn't go as planned. All I could do was say the few words I had memorized and hope for the best.

Opening my mouth, my heart was racing as I said, "Anna, mi amo Stefania, posso matrimonio tua filia" (I love Stefania, can I marry your daughter)? Flush with embarrassment, it was almost incomprehensible to my mind what I'd just done. Gasping for breath, I watched as Anna slowly looked up from the orange. Waiting, I questioned if I'd said the right words or done the right thing. With no response yet, I didn't think she understood, so I decided to repeat what I said slowly, more clearly, and more punctuated, *"POSSO. MATRIMONIO. TUA. FILIA."*

Confused, Anna sat quietly. Now, it became a little awkward. I was anxious, and I had no idea whether or not she knew what I was saying. Starting to repeat myself a third time, she interrupted me, saying, "Aspetta" (wait). Then, she picked up the phone to make a call.

With the ring echoing into the kitchen, she put the phone on speaker while it rang. Shrinking into my seat, I now felt somewhat like a fool. I was clueless about the appropriate way to propose, and I had been afraid to talk to Stefania directly, wanting instead to surprise her. Unable to go forward or back, I just sat there and listened. Then, I heard the word "pronto," and my heart sank. On the other end of the phone was Stefania answering the call. Mortified, there was nothing I could do but listen to her talk to her mom and wait to see what was said.

It is worth noting at this point in the story that one of the most interesting parts of learning a new language is that you comprehend what you hear at a much faster rate than what you speak. It seems the brain has different centers for comprehension and conversation, and my listening center knew some 60 percent more of what people were saying than my speaking center. So, listening to the conversation, I understood that Anna was telling Stefania exactly what I said, and she was trying to figure out why I was asking her.

Overwhelmed with embarrassment, I had not intended to say anything to Stefania until I proposed, which, in hindsight, was probably not good. But now at this moment, I didn't know what to say. Passing the phone to me, Anna said, "Parla" (speak), and so I did. I explained the situation—that I was trying to ask her mom permission to get married. Stefania now was even shocked. This was not an Italian tradition, and she began to laugh, which was more embarrassing. "Why are you asking my mom?" she said. Then, surprised and having registered the situation, she followed up with, "You want to get married?"

At a loss for words, I decided it was best to talk later and passed the phone back to her mom. Stefania translated, and both of them laughed with some more glee at my expense. My romantic plan was over. My cover had been blown. Now, there would be no surprise as to "if" I would ask, just when.

VIVE PARIS

Having suffered through the embarrassment of asking permission to marry Stefania and having bought a very basic engagement ring in Rome some six weeks earlier, I was now on phase two of my plan, the proposal phase to take place in Paris, and I had hoped that it would go better than what I had just experienced.

With the skidding of wheels on the tarmac kicking up dust, we arrived and skipped down into the subway to get to where we would be staying at a hotel just down the street from Monument a la Republica. This was the first time I had been to Paris, and the romance of the place was palpable. Just below on the first floor of the hotel, you could smell the coffee brewing and pastries being made as glasses and silverware clinked into place for each customer. Buzzing past and honking was the sound of taxis, and there was a hum in the air outside on the city streets. It was true what they said; this place was special.

Underfoot, I could feel the cobblestone of the walkways as we left to take in the first sights and smells of the city. In front of me was the towering Gothic architecture of Notre Dame with its pointed arches, ribbed vaults, and flying buttresses. As jaw-dropping as it was alone, against the backdrop of the city, it was mesmerizing. I felt as if I was in a scene from *Les Misérables,* the story of the rebellion of 1832. Walking with Stefania along the encased walls of the riverfront, I reimagined the scene where Inspector Javert finally ends his chase of Jean Valjean and falls backward into the Rives de Seine (banks of the Seine), killing himself in a form of self-redemption. It is a strange scene in a powerful play and all the more impactful being here in this romantic city with its Renaissance beauty mixed with the Haussmannian architecture.

Entering The Louvre, the famous French museum that possessed the most famous of paintings, I could feel the silent energy of the space as we moved from room to room, observing visions that could travel you back in time. Entering a giant square room in the center of the museum, I waited in the switchback line of turns full of people to see what seemed like the most basic of paintings but was actually the most legendary. It was the *Mona Lisa* herself. There, I paused to grasp her energy and comprehend her hypnotic gaze. Passing slowly, I watched while her eyes seemed to follow only me across the room from right to left. It was a grand moment in one of the most romantic days of my life, and soon it was to get more romantic at our last stop, the Eiffel Tower.

Biting deep into a fresh baguette with prosciutto, we slowly migrated through the city to our destination, tasting and touching a little of everything. At the same time, a vision of my cause, the proposal, occupied my mind, and I could feel fear creeping up beneath.

In the foggy mist of the day, I could see rising from the center of Champ de Mars the dark steel silhouette of the Eiffel Tower. Our feet were worn with exhaustion, and you could feel the miles we'd covered in our clothes. And for me, the years of failed relationships and failed attempts at love were creeping into my head.

CHAPTER 10

Behind them came a cloud of anxiety and self-doubt, which were doing their best to stop me. This was a big decision I was about to make, the biggest in my life, and I was doing my best to hold my emotions together.

Inside, I began to ask myself, *Was I really ready for this? Was this the right time? Had I given this thing enough thought?* In every case, the answers were unclear, fogged by emotion, but what was clear was my direction forward. I had known for a long time that this was what I wanted. I had been building toward it at every move. Now, in the chaos of execution, I was doing my best to hold onto my plan and continue forward.

Reaching deep into my pocket, I nervously wrapped my closed fist around the ring I'd bought and squeezed it tight. Being in love and without a care in the world, I'd bought the ring with my last paycheck, knowing full well that if I didn't get my passport renewed, I'd not be going back to work or most likely not staying in Italy.

Stepping into the space of the Eiffel Tower, I was getting closer and could feel the groans of the steel columns holding up this massive structure. From its base on the ground to the Premier Étage (first floor), there were four enormous legs supporting the giant platform standing 187 feet above the ground. After purchasing a ticket, I could hear the footsteps and echoes of foreigners from around the world, which distracted me from the cause at hand—my proposal—looking forward to some of the best panoramic views of Paris.

Reaching the first stage, we wandered around the gift shop and restaurant and then toward the elevators that would take us to the Sommet (top), where I would execute my plan. In somewhat of a panic, gravity had seemed to increase tenfold, and every step felt harder than the last. Reaching the elevator, I went to push the button UP when I noticed a large sign in front of me across the entrance, saying, "LE SOMMET DE LA TOUR EST FERMÉ POUR RÉNOVATION." Pushing in closer, I squinted to read the English text in the smaller font: "The top of the tower is closed for renovation." Confused about what to do next, I could feel my fist open, releasing the ring inside my pocket. With all my mental preparation, I had not planned for this. Turning away sad, I was at a loss for what to do next.

Leaving the Tower, we dragged ourselves across the cold and rainy streets. Now, Paris somehow felt more depressing than romantic. Having been caught off guard, I lost focus and began to look for a resolution. Looking for a new place to ask, I looked at parks, riverfronts, and outlooks, but none seemed to grab my eye. Paranoia began to take over, and I began to ask myself if this was all just meant to be.

Arriving back at our hotel, I took out the oversized room key, rattled, and unlatched the door, then pressed inside. As Stefania was tired from the walk, she laid down, and I decided to take a shower to wake myself up. Halfway between imagination and reality, lost in a deluge of defeat, I noticed things in the room I hadn't noticed before. The walls had cracks running down the opaque pink paint, the bathroom tiles seemed to be falling apart in slow motion, and the shower worked more like a fire hose than a shower. As steam filled up the small space, I stepped under and through.

The time we had spent together came to mind, and I began to envision myself as an actor in a scene of my own movie, trying to figure out if I was going to end the scene a hero or a zero. In our time, I'd found Stefania to be like no other woman I'd ever met. When I arrived at her doorstep empty-handed, she'd all but given up her life to be together with me, this random man from across the sea. Adding to that, her family had also accepted me. From day one, when I met her on that barren beach in Costa Rica until now, my instincts told me that this woman was the right woman for me.

With a twist, I shut off the water, reached for a towel, and wrapped it around my waist. The hot shower had helped push aside my fear. I knew now what I had to do. Half-naked, dripping wet, and with my right hand grasping my towel, I grabbed the ring and resolutely stepped out of the bathroom—there in the most unassuming of places wrapped in the romance of a poor French hotel, I got down on one knee to propose.

CHAPTER 10

Stefania, still recovering from her nap, sat on the side of the bed in shock. As she looked at me with some amusement, she worked to figure out what was happening. There and then, I looked her right in the eye and said, "Amore, will you marry me?"

I could see that she was baffled. For her, it seemed as crazy as anything I'd ever done. Like the strangely romantic meeting on a beach, chasing her to Italy, or moving to this foreign land, this was just another page in a book she was falling in love with. As she continued to mull it over, I held my breath for what seemed like a lifetime. And then, in a blessed moment of relief, she declared, "Yes!" And with that, I gasped for air; we were getting married.

Shooting recreations in India on the road up to Tanglang La. In front, James Herron, the director of photography, is walking with a camera, and behind, Barry is dressed as runner Mark Cockbain (2012).

"Before enlightenment; chop wood, carry water. After enlightenment; chop wood, carry water."

—*Zen Buddhist saying*

CHAPTER 11

ULTRA-RUNNING, THE HIMALAYAS, AND THE HIGH

ALBANIA, ITALIAN CUSTOMS, AND MARRIAGE

Landing back in Rome with my new fiancée, I was proud of having accomplished my primary mission. Now, I only needed to get a stamp on my passport, and I'd be golden for six more months. Grabbing our bags off the carousel, we headed for the exit to what I believed would be customs. Walking out past the ticket counter, we exited the airport into the open space of the parking lot, then came to the painful realization that France was not considered a foreign country, and there would be no stamp from customs.

Returning home, I did the research that I should have done before leaving. Clicking through Wikipedia, I uncovered the essential details of the Schengen Agreement in 1985 established to improve trade among a large group of European countries. Its creation led to the abolition of international border checks, making it easy for people to move fluidly between countries. For the larger economy to compete with major countries like the United States, this was a good move, but for me, it was catastrophic. Now, I would need to pick a country outside the treaty nations and fly there fast.

Digging deeper, I realized that the nearest country outside of the Schengen Agreement was Albania—a former Communist country that had undergone the loss of private property rights and religion in 1944 after World War II. It had also suffered poverty and struggles, which historically followed all Marxist-Communist regimes. It remained frozen from economic growth until freed from oppression in its Parliamentary Elections in 1992. After that, it became a free market independent state and could be accessed by flying directly across the Adriatic Sea. Locating some family friends living in Albania, I bought a ticket and made plans to fly solo and meet them there for a few days. Taking off from Fiumicino Airport, I made the hop over international waters to the "foreign" land of Albania. A stamp was soon to follow.

Getting picked up by my Albanian hosts, I jumped in the worn-down classic Russian Zaporozhets ZAZ-965 model car. I squeezed my long American frame into it, finding space in the back seat. After making our obligatory greetings, we headed to a local restaurant, which felt like the living room of a large family dining room on a foothill overlooking nearby hills rolling to the sea. After settling down, the waiter brought out the traditional Albanian dish Tavë Kosi, a baked lamb and rice casserole, sizzling on a hot plate with a yogurt topping. My hosts then pulled out a plastic bottle and poured a hearty amount of homemade Apple Grappa, a sort of vodka. With the purifying heat of a fresh pour, we cheered across the table, threw it back, and then I let out a dry, hard cough.

Exiting together, we stepped out into the cool, dry air of the capital city of Tirana and began to walk the streets near our hotel. Entering Skanderbeg Square, I could not help but notice that there was still a lot of state-sponsored, centralized architecture called the Soviet Bloc left standing from the communist era. That redundant military energy could be felt in viewing the statues of

CHAPTER 11

heroes on horses and the rigid iconic columns of buildings that surrounded the bazaar. It felt like its own version of a cathedral honoring a religion but the worst kind—the kind with no God. While I had my issues with the institution of the church and its rigid ideas of God, this form was far worse. In this "religion," authoritarianism reigned supreme and bred nihilism and a lack of hope in the population. Hope comes from faith in the future that is provided by both the freed human spirit and the belief in something greater than ourselves. In this place, before the fall of communism, the iconic idea of "man" stood at the top of the human moral code. Looking around, I loved the stoic nature of the structures but hated what they represented. I believed they should all be torn down.

Waking up the next morning, I sat down at a breakfast house to eat. In front of me was a flat fried egg that had the texture of a rubber chicken. Taking a bite, I worked my way through some of the saltiest, most over-cooked eggs I'd ever eaten. Finishing up my plate, I wrapped up my twenty-four-hour excursion with my hosts picking me up as we toured the local shopping mall. The structure reminded me of what I'd have in my hometown. While it wasn't my dream to tour the place, my Albanian hosts shared it with pride as if it were an extension of American culture. And it made sense later that this emblematic place was a seedling of capitalism and the end of the oppressive regime.

Departing through airport security, I held a small Albanian flag in my hand as I waved goodbye to my friendly hosts, then jumped on a plane back to Italy. Touching down, I walked up to the border agent and turned over my passport. Watching him look over my record behind a thin veneer of plexiglass on the digital screen, I was excited to finally be getting it stamped and going on with life. But life was slowing to a tick as the agent continued to comb my details for an unusually long time. Breaking the silence, he turned and looked me up and down. Then, in Italian, he picked up a radio, clicked the two-way receiver, and made a call to another agent. In all my times walking through customs, I've never had to stop and wait, so I was arguably confused by the skirmish.

In the distance, I looked to see the other agent, clean-cut and dressed in uniform, waving me over. The guard at the desk then escorted me into a back room. I could feel them watching as they closed the door behind me and sat me down in a waiting area where there were a few small benches. It had the feel of a precinct and the pace of a government office—clearly with no concern for my time. Twenty minutes or so later, a well-dressed agent walked up and directed me to another smaller room. There, under a long fluorescent ceiling light stood the Gestapo—a big man with a mustache, glasses, and a long face. Sitting behind a desk with random stacks of paper, he leaned forward and started to interrogate me about my travel. I had seen this stuff in movies and even imagined myself being like James Bond escaping the officials trap, but being here in real life was slightly different, and I was working to get my bearings so I could gauge the seriousness of the situation.

Shifting side to side in the steel frame chair, I was adjusting myself to get comfortable while being asked over and over again in English with an Italian accent, "Why do you go to Tirana; what were you doing in this country for such a short time, and what do you bring?"

Calm for the situation, I gave enough info not to alarm suspicion but not too much as to give myself up—"I am a tourist. I was curious about this country and had some friends there to visit."

Time after time, they asked the questions, looking for a different answer. But I knew I had done nothing wrong. I wasn't a smuggler of any sort and was only getting my passport stamped. In reality, I wasn't sure if this was punishable. I didn't know if I'd get sent back to the United States and rejected to enter, which would be horrible—an end to the momentum that I'd been building with Stefania. So, I continued to answer questions calmly, and they continued to ask.

After about an hour of interrogation, they finally gave up and led me back to the waiting area. I pulled out my flip phone and quickly called Stefania to tell her the details. She had been waiting in the parking lot to pick me up and was in a panic. She, too, was concerned that the Italian government would deport

CHAPTER 11

me. We were not in America, I was not a citizen, and the rules that apply to citizenship did not apply to me.

While I was talking to Stefania, the man who had been behind the desk stepped back into the waiting room, so I needed to hang up. Leaning forward, he called me back into the office. I was now beginning to feel like a criminal and wasn't sure if I should laugh, cry, or just tell the truth. However, remaining in character, I answered every question the same as previously, and they repeated them again and again. But this time, they interspersed conversation to break up the mood and catch me off guard. Finally, after nearly two hours, they decided to let me go. However, as I walked toward the exit and freedom with the same passport in hand that I'd have to show my employer to continue to work, I suddenly remembered I had not gotten it stamped.

As one of the guards ushered me toward the exit, I knew that if I walked out that door and into the state of Italy, I would not be getting a stamp—possibly ever. However, if I paused and asked for a stamp, the whole episode could start over again. But I had no alternative but to stop. Turning to the guard, I said, "You forgot to stamp my passport, and I don't want to be unlawful and leave without a stamp." Smiling at my desired compliance and somewhat confused, the man pulled a stamp and ink pad from his desk. With a solid boom on the desk, he stamped my passport and dismissed me. Finally, I had six more months to figure out how I was going to live in Italy.

Back at home, we started talking about plans for a wedding. Having overcome some serious fears of commitment and having proposed, I was feeling better about myself and more confident in my ability to make a big decision and follow through. Given our most recent challenges with getting a renewed tourist visa, we decided it would behoove us to have a legal wedding sooner rather than later. Organizing things through the state, we were set to get married in a 1500-year-old building owned by the state of Italy just a few blocks from the Colosseum. Inside, the place felt like an extension of the walls of Rome. Built with ancient bricks, which, with age, seemed to have turned into stone, you felt completely encased in the arches and vaults holding up the walls and the dome ceiling above. There, on the day of our legal wedding, we gathered together with a small group of our closest family and friends. And in the warmth and intimacy of that enchanting historic space, we were ceremoniously married, stating the traditional "I do" before being presented with a giant, thick, granular paper certificate to sign. Pressing down on it with a classic feathered ink pen, I wrote my name, followed by Stefania, who did the same. We were now legally married, having romantically done so in the eternal city of Rome. I could hardly believe it.

A NEW FILM AND THE HIGHEST RACE ON EARTH (2010)

In the mix of my marriage and move, I was starting to reap some rewards from my film work. My first documentary, *Profiling HURT,* was now being sold and circulating via DVD sales. Thrilled, I had finally gotten a small taste of success and wanted more. I wanted to do something bigger that would earn the respect of my peers. I needed to find a project that would catch people's eyes and further solidify my work. No sooner had I started thinking that than a new project showed up.

Scrolling the classic feed on Facebook, I noted that I'd gotten a message from my friend Dan Bree. Dan was a fellow television producer whom I'd known from my days working in San Francisco. He had heard about my recent UltraRunning project and was calling out a post regarding a race in the Himalayas. In that post, the race director had written that he was seeking "journalists and filmmakers." Following up on the lead, I replied to the post and connected with the race director, Rajat Chauhan.

Rajat Chauhan was from a family in the upper cast of Indian society and was now a doctor performing athletic medicine in a clinic in New Delhi. During his time in medical school at the University of Cambridge in the United Kingdom, he had picked up UltraRunning. Obsessed with his love of long distances, he ran the Paris-to-London UltraMarathon, among others, before graduating from medical school and returning home to New

CHAPTER 11

Delhi. Having grown up going to a private school in the foothills of the Himalayas, he had decided to pioneer one of the first UltraRuns in India over two of the highest motorable passes in the world.

The physical race would be run in the region of Jammu-Kashmir in northern India between the borders of Pakistan and China. To complete the challenge, runners would need to cross 133 miles (222 kilometers) over—Khardung La (18,300 feet) and Tanglang La (17,500 feet)—before finishing in the high desert of the Nuru Valley at 14,000 feet. In addition to the altitude, the runners would have to survive the high desert heat, the high mountain cold, dehydration, exhaustion, and the dangers of high altitude. Reading the post, it sounded completely insane, which I loved, and I wanted to learn more.

Like so many other times in my life, this moment felt somewhat divine and called out to me. I had just finished my first film on the topic of UltraRunning, a niche sport, and knew exactly what I was up against. I also was looking for a bigger story to tell, and this was definitely bigger. Chasing down some details, I found Rajat's contact info and reached out directly. Shortly after, I heard back, and within a week, we were jumping on a Skype call and making plans for how we might be able to pull this off.

When we connected, I asked him a few initial questions about what he needed, and then we had a long detailed conversation on the run. With enthusiasm and passion, he shared this would be the very first time anyone had run the race. He was working hard to recruit press and runners to help pioneer this event and tell the story. He currently had hundreds of people responding with interest, but only a few seemed serious. Intrigued, I asked him to systematically explain the course so that I could properly understand what I would be getting into.

Reimagining his experience running, driving, and mapping out sections, he shared that to do this either as an athlete or a journalist, we would first need to fly to New Delhi, where everyone would initially meet. Then, we would travel by plane to the city of Leh in the region of Ladakh, south of Jammu Kashmir.

People would prep there in Leh for the altitude before ultimately being shuttled two hours away to the starting line.

Leh, the host city, is known as the highest city in the world and rests in the valley at 11,000 feet between the Stok Kangri Range to the south and the Ladakh Range to the north. A little haven surrounded by giant peaks, the geography is rugged and arid, and what little vegetation they have can be found to have sprouted from the glacial-fed Indus River. To help put the altitude in perspective, compare it to the Rocky Mountains Range in Colorado, where the highest peaks top out at just above 14,000 feet, and most are around 12,000 feet. We would be landing and sleeping in a village where the sidewalks and hotels were at the altitude where most mountaineers in the US spend days working to reach.

To travel there and then traverse higher, Rajat informed me that everyone would first need to prep their bodies for five to seven days to become acclimated to the high altitude. This was a mandatory requirement made by the Indian military, and for good reason, too. There had been multiple cases of travelers flying up from sea level on the same day and landing in Leh, then renting a car and driving up to Khardung La, only to overstay their welcome and end up getting sick, hospitalized, or in the worst case, die. That is dead from untreated High Altitude Pulmonary Edema (HAPE) or High Altitude Cerebral Edema (HACE)—hard to believe but true. For Rajat to have gotten permission from the military, which oversees all activity in the region, everyone needed to acclimatize.

So, the prior days spent preparing for the race were mandatory and designed to be able to get the runners acclimated, a process of increasing the number of red blood cells in the body for better processing of the reduced percentage of oxygen in the blood. To do so, they would take walks each day—first at 13k, then 15k, and up to 18k, where they would first see Khardung La. There they would walk around the Stupa (structures built by the local Buddhists as a place of meditation), stand under the colorful prayer flags stretched across like clothing on a line, and have a cup of butter tea and biscuits from the café on top

CHAPTER 11

of the mountain pass. Once they'd arrived at this point, Rajat explained, they were ready for the long, gruesome walk-run up and over 133 miles, crossing both Khardung La and Tanglang La.

With the audio of our conversation traveling from India to my home in Italy, I listened on Skype to Rajat's description of the run, I was dumbfounded. This race felt like an insane feat to run, let alone film, and would be an incredible story to tell. Building with excitement, I realized that I was one of the first people at the gate. This was my story for the telling. However, I also comprehended that this would take time, preparation, and thought for me to execute. For Rajat, who wanted me on a plane tomorrow, there was not enough time. It was July once we'd connected, and the race was starting in August, just over a month away. So far, the first three of his runners had committed—two from the United States and one from the United Kingdom. They had signed up, purchased tickets, and were preparing to pioneer the first attempt. Selling me on the idea, he shared that if I wanted to capture this, I needed to make it there immediately.

Closing the Skype portal, I began mulling over the details of how to produce this. This project would be a giant undertaking. I wanted to jump and go, but making it happen would be hard. Taking it all into consideration, I wasn't afraid. In fact, I was confident that if I did do it, I would execute on it and return with great footage. This confidence was a new achievement in both my professional and personal understanding of what it took to execute one's dreams. However, having now overcome fear-driven decisions and impulsivity, I needed to apply clarity and logic to figure out the best plan of action for achieving this feat.

The first step in this new process of gaining clarity and making decisions based on intent over impulse was to meditate and pray. I needed to breathe into the moment. Doing so calms the mind and centers the spirit. It helps you put off the impulsive nature of emotional decision-making and focus instead on your intended outcome. Here, I begin to see clearly what I want through my frontal lobe and activate the pineal gland, also known as the third eye.

The next step in the process is prayer. Prayer recognizes a relationship with something greater than yourself and accepts that there is an unknown to all decisions. By praying, you are humbling yourself in the request for help from your Creator. By sacrificing your sacred time, accepting that there are things you cannot control and shifting your mind from the self-centered to the divine, you can access a relationship that would otherwise be left untapped. In this moment, you can get clear on those things you can and cannot control. As I begin to pray, I request guidance for a path forward in my decision-making. I ask for help with the things that feel bigger than myself to take on. And I turn a portion of the new dream—in this case, the dream of filming this race—over to something greater than myself, God. In return, I feel a release from a desire to do this on my own with the sense that the path forward will be, in part, the small actions I need to take and in part being open to the things I cannot control.

Using this newfound skill, it was now clear to me that I would make this documentary happen, but I needed to wait until I was properly prepared to go and film. Reaching out to Rajat again, I informed him of my decision to wait till the following year, knowing that if he didn't complete the race, it may very well never happen. Then, I went on to work out the details of how to get there by August 2011 as he went off to pioneer his very first race in 2010.

LA FAMIGLIA AND "THE MAFIA"

After the fiasco of returning from Albania to Italy and having gotten legally married, I was now able to apply for a Permesso di Soggiorno, an Italian residence permit, which would allow me to attain a work visa. Italy was now becoming a home for me, and the family was opening up more and accepting me in. In the act of getting married, I earned a newfound respect. I was now on the inside of Italian culture in a way that I had never been before. What that meant was that I would be taken care of. I would always have food, shelter, and enough to survive, and

CHAPTER 11

no one, no matter how hard times got, would let me go hungry. To use a throwaway line from *The Goodfellas,* I was a "made man," but it was more than a throwaway line in an American film, much more.

During the past year of my life, I'd also become friends with Angelo, the long-term companion of Stefania's mom, Anna. Angelo was like a pseudo-father to the family and had made an impression on me from the first time we'd met. When I arrived in Italy over a year ago, I was somewhat like a stray dog, lost, and Angelo instantly looked after me. The first time we met, it was a cold day in Rome. I had only arrived and was working to put order to things. Sitting at the table in Anna's kitchen, I only had a T-shirt on and was chilled to the bone on a damp, cool day. Walking into the house, Angelo shouted out, "Americano!" Turning to me, he could see that I was chilly. After only knowing me for a matter of minutes, he took off the warm winter vest that he'd been wearing and gave it to me to keep me warm. I felt like a little puppy dog getting a treat from his master, and instantly, I was enamored and fell in love.

For him, I was the Americano whom he liked to call "John," which is a known classically overused American name. To me, he was a version of the *Godfather,* who's always there to take care of the family in our time of need. Since I'd been living near Anna, he started to randomly stop by our apartment down the street to bring over bags of groceries or coupons for gas. Often, he'd call us impulsively, then grab Anna and take us all out to eat at a restaurant of a friend. It was never just a restaurant; it was always a special seat and always people he knew. To help with work, he'd donated a scooter for me to drive. Feeling like a kid, I used the new ride to race around town until it broke down. Never shy with his money, always willing to give, and never asking for anything in return, Angelo had become a sort of hero.

On occasion, when I was home from work, he'd give Anna a call and tell me to meet him out in front of the apartment. After his arrival, he would drive me around from place to place in one of his variety of small Italian cars. He owned twenty cars ranging from a vintage 1960s Alfa Romeo Spider to a Fiat

501, the equivalent of the Model-T, which was used to chauffeur Stefania to our wedding. For dinner, it was most often his Cadillac Fleetwood, or if it was for work, it was his Fiat Veicoli Industriali (lorry tow truck). But the one I liked the most was his 1989 Fiat Cinquecento, a small, economical little car that Angelo seemed to use to hide from all his wealth and success. The two of us would tool around town—the millionaire dressed like a dirty garage mechanic and the American. Nobody knew the better.

While gripping the wheel with his thick sausage-like fingers, Angelo would get great joy teaching me all the Italian swear words. Waving his hands around in violent motions, he would turn to me and say, "Ripetere, CHE CAZZO DICE!"

Then, on cue, I would repeat, "CHE CAZZO DICE."

Instantly, he would break out in mad laughter, knowing I had yelled out "what an asshole" at the top of my lungs! It was a unique time to say the least, and I loved to see him laugh.

A day after the wedding, I'd gotten a call to meet him in front of my apartment at 6:00 a.m. Jumping out of bed with sleep still in my eyes, I raced around to get ready and then waited at the gate for his arrival. Pulling up in his tow truck, I jumped in, and we headed off together. On this day, he wanted to take me on a tour of all his businesses and introduce me to some of his esteemed colleagues.

One after another, we would go into cafés, restaurants, construction sites, and tow yards to meet people. At each location, people would respond, saluting Angelo by name. At one bar, I laughed as he reached out his right hand full of rings toward the bartender who ran the joint. Without pause, the guy leaned forward and kissed the ring, literally. I couldn't tell for certain if it was a gage for show or the real thing, but I understood that since Angelo owned the restaurant, there was possibly more to it than met the eye.

From day one, the observer in me was caught up in the character study of Angelo. As the filmmaker, I liked to watch his gait as he walked. He'd push his shoulders back, prop out his chest, and boldly take large steps at a quick pace. He always appeared larger than life, and because of this, his influence over

CHAPTER 11

people was surprisingly powerful. He was able to tame the most dangerous-looking individuals and seduce the most intelligent of women. On occasion, while I was riding shotgun, he would ramble on in Italian, claiming that his connections went all the way to the Vatican. "Ho lavorato per il papa" (I worked for the Pope), he would say. While I had no confirmation of that, what I did understand was that Angelo had built his way up from the bottom to respectable wealth, and in Italy, that isn't easy and comes at a cost. You'd never know it at first sight; he always wore laymen's clothes and worked hard labor day in and day out, but Angelo had a secret all to his own.

By late afternoon, we'd toured several locations throughout Rome, and his businesses seemed to be nearly endless in scope. After hours of driving, we were back close to where we lived and had one final stop. Pulling into a big parking lot, we arrived at a large set of tall warehouse buildings. Listening to the gravel crunch underfoot, I watched a group of six guys jump off forklifts, everyone stopping what they were doing to greet Angelo. There, in a dirty supply yard filled with cars and pallets, Angelo was in the center of a semi-circle regaled by the men.

For the first time in all of our previous visits, Angelo introduced me to the group. I was standing just behind him and off to the right. Like a bodyguard, I was just there to listen with little to say. The first words I heard and understood were, "Americano," saying, "È il marito di Stefania" (the husband to Stefania).

Then, he paused to offer something that felt like it was right out of a movie. In his Italian slang, I could hear him say, "Lui e un amico mio" (he is a friend of mine).

Looking around at the faces, I nodded at the men who recognized me for the first time and gave me their respect.

Amid the conversation, I noticed a man I'd seen before. His name was Giuseppe; he was the checkout clerk I'd greeted at the hardware store near our house. I had gone there a couple of times in the past to buy odds and ends. I wasn't sure if he'd seen me, but I knew him and made note of his presence.

About a week later, after riding with Angelo, I returned to the hardware store. After I finished grabbing a few items, I walked up to the checkout counter to be rung up. Pausing, I looked at Giuseppe, who smiled at me and then glanced at the items. Wiping his hands back and forth as if to brush off the dust, he looked me directly in the eye and said, "Niente." Surprised, I was pretty sure he was suggesting that I didn't have to pay. As foreign as it was to me, it suddenly became clear that in Italy, there was another currency that surpassed money. That was the currency of being connected. Grabbing my stuff, I walked out and realized for the first time I was truly in the family.

THE AUSSIE CREW AND THE COLOSSAL FAIL AT CUSTOMS

Over the same window of time, I had also been developing my plans to film in India for the 2011 race. Talking to Rajat every month since we'd connected, I'd learned that the first running of his race in the Himalayas had been a success. While two of the runners had dropped out, they had a finisher, which made it official that it could be done. Plans were now in place to have the second run in August 2011. Excited to hear of this success, I continued forward with my plans to travel to India and capture the race. For travel and expense, I decided to use funds from the DVD sales of my last project, which was enough to cover my costs. For the production, however, I would need more than myself. I needed a team to capture this moving three-day story in some of the roughest countryside in the world. When meeting a young filmmaker named Cody through a common friend in Italy, I pitched the idea, and he was on board. Being from Australia, we started to communicate via Skype and make plans. Over months of planning, he'd also recruited a friend, and the two of them were bringing the best in new HD cameras for the race. This was awesome, I was finally legit—no longer just a solo guy with a camera—I now had two professional filmmakers on the crew. With just a few weeks to go before departure, all the logistics were in place, and the guys were waiting for their visas to be approved for travel. Then, the worst that could happen happened.

CHAPTER 11

While working to process visas for the trip, the two young Aussie filmmakers who'd never done this type of thing before decided to fill out the applications for WORK VISAS instead of TOURIST VISAS. I remember some months back on a Skype call discussing the topic and hearing the concern about their gear getting through customs, but at the time, I recommended going as a tourist and not making a big fuss. Concerned for equipment, they disregarded my advice and decided to go the route of applying for a WORK VISA. Then, just a few days before their departure date, they got word from the Indian Embassy. In an email from the office of the consulate, they read that the applications had been REJECTED. Then, with little time to self-correct, they started to scramble for Plan B.

Listening to them share the news, I was in complete shock. In an instant, my heart sank, and I began to slouch into my chair. If there was one thing that I knew, it was the failure of institutions to be able to self-correct in short order, even worse, bureaucratic institutions. Nearly in tears and still in denial, I knew we were in trouble. Ten days out, the boys sent emails to the embassy requesting a change to the classification from "work" to "tourist.' Five days out, they made calls. Three days out, they drove down from the Gold Coast to Sydney to visit the Indian Embassy in person, where they called once more, this time from the embassy lobby. However, once again they were rejected, meaning that the trip was over for them.

By the time I'd heard the final news, I had already assumed the worst and was working to make sense of my luck. I'd spent a year planning every aspect of this production and, just a few days before departure, was without a crew or cameras. Looking at my little DSLR Panasonic GH2 camera, which had been meant for behind-the-scenes, my heart sank. My face turned flush with shame. Failure again was raring its evil head. It felt like I'd shown up just to prove that I wasn't up to the job. Then, with the encouragement of my wife and a call to Rajat, I started to see this differently. Instead of failure, this was a test and a trial of sorts, one I would not lose.

Having alerted Rajat to the challenge, he went to work sourcing local filmmakers in India while I got on a plane for the eighteen-hour flight.

NEW DELHI, LEH, AND ACCLIMATIZING IN THE HIMALAYAS

Arriving on the streets of New Delhi was a total culture shock. I'd traveled extensively throughout second and third-world countries, but the poverty at this scale was immense and hard to grasp. People were living in overwhelming numbers under bridges, in broken-down buildings, and on rooftops with little more than their sleeping mats. Many streets were poorly maintained and sometimes dropped off quickly into dirt paths with trash strewn about. Cows could be found wandering along sidewalks and across busy streets, leaving random cow turds like land mines needing to be missed. Bustling traffic and incessant horn beeping ruled, and there was a constant smell of burnt trash in the air. Yet, against this backdrop, there was a mystical feel to the place that I could not put my finger on—something that felt both very real and fictional at the same time.

With the dirt road below and in the sweltering heat of an oven above, I wandered the streets near our hotel and felt like I was a time traveler. On multiple lanes of any given road, I watched as the most modern of rides would bob and weave between a collage of vehicles ranging from a tuk-tuk or motorized rickshaw, a random ox or camel pulling a load of dirt on a homemade cart, and a family of three on a scooter. In the midst of this chaotic race for life, you'd find flames rising out the sides of caldrons of food and barbers giving roadside morning shaves on the sidewalk. At times, it felt like this: if you had a lightsaber and a Landspeeder, you'd be in a scene right out of *Star Wars*.

While it would be easy to dismiss this poverty and chaos as the fault of the culture, you'd be wrong. While I did not understand every factor going into the makeup of India, I did understand that the work ethic and drive of this culture were like nothing

CHAPTER 11

I'd ever seen. For me, the best example of this was the doorman at my hotel. Getting settled, I'd found that he worked and slept twenty-four hours a day behind the desk. I first made note of this when I was dropped off by taxi from the airport at 3 a.m. and he popped up from his mat to serve me. When I woke up the next morning at 9 a.m., he pointed me to the coffee, and when I returned the following evening at 6 p.m., after a day touring the city, he welcomed me back and gave me a mint without missing a beat. Only then did I find collectively throughout India how people in this culture pushed themselves harder than any culture I'd ever known.

After a day in the country, I connected with Rajat and quickly learned that he was hardworking and extremely resourceful too. In the short time between my departure from Italy and my arrival in India, he had connected me with two local Indian filmmakers, and we now had a new crew in place to help. Meeting up at his office in New Delhi, we discussed our plans for travel to the Himalayas and the execution of production at high altitude. Having gotten all the runners into the country, we soon jumped on a plane from New Delhi and flew over mountains to the high country and the city of Leh, where we'd begin the process of acclimatization.

Flying out of New Delhi, I watched as the landscape changed from the hot and humid sea-level jungle to the snow-capped peaks of the Himalayas. Gazing out across the mountain range, I was mesmerized by the idea that somewhere out there was Mount Everest, the presumed "top of the world." I pressed my head into the small window of the plane to see if maybe I could see it. With nothing but glacial valleys below and snow-capped peaks as far as you could see, we started to descend, and I now looked down on the golden-brown desert surrounding the emerald Leh Valley below. There in the center of all the sandy brown was an oasis of green land covered with rice paddies. In the middle of all that green was a cluster of buildings and roads that was the city of Leh.

Leh and the surrounding area evolved from an ancient culture dating back thousands of years and was an important stop on the historic Silk Route connecting Central Asia with South Asia. Harnessing the flowing waters coming down off the glaciers above the pioneers of this land are believed to have slowly fertilized the brown lifeless dirt with organic and human waste, making it fertile. By and by, the land seeded fields, fields seeded rice paddies, and the paddies fed the growing community. Amidst all this, Buddhist philosophy took hold in the third century AD, and for centuries, the people in these mountains lived a mostly secluded life away from the influences of the modern world. Then, after the advent of the airplane, transportation increased, and people began to visit more and more. Today, Leh has undergone some substantial growth. However, you can still see much of the ancient remains of monasteries. Due to the hard climate, the culture remains mostly intact, making it a rare place to peek into a way of life mostly absent from the rest of the world.

Exiting the plane, we walked directly onto the high-altitude tarmac, and I could feel the cold fresh air against my skin. A part of me felt like I literally expanded in this space—as if my spirit felt comfortable enough to spread out and take up more room. Walking into the terminal, I could hear the drum beats of the damaru, a two-sided drum that monks use to create a rhythmic tone. We were surprised by a greeting from Himalayan monks who welcomed us. They also placed a khata or Buddhist scarf around our shoulders as a gesture of goodwill.

Having been warned of the need to go slow and acclimatize, I had expected the altitude to hit me like a harsh wind, but I had not even noticed it until arriving at my hotel. With bags and camera in hand, I started up the one flight of stairs and suddenly felt dizzy, my head spinning, and I needed to stop and catch my breath before I passed out. Grabbing onto the wooden railing running along the stairwell, I squeezed hard, making a tight fist, and paused to inhale. Recovering my breath, I walked to my room, quickly unlocked the door, and sat down on the bed. I had done my share of mountaineering summiting Mt. Whitney (14,505 ft), Mt. Shasta (14,174 ft), and others, but never had I landed at such a high altitude without acclimatization, and I knew this stuff was not to be trifled with.

CHAPTER 11

Looking out the window from my room, I could see a small village of cars and people moving in every direction all around me. Life here seemed as normal as anywhere. Yet, I knew in my head we were at 11,000 feet above sea level. In my history of climbing, I'd spend nearly two days working hard to reach this altitude on foot. But now, having arrived in just a few hours by plane, I knew I needed to be extra careful.

Taking deep breaths, I had laid down to rest when Rajat stopped in to inform me that one of the Indian crew members he'd recruited to assist me was sick. He had been vomiting and struggling with his breathing. A medical team Rajat had organized for the event diagnosed that he was struggling with High Altitude Cerebral Edema (HACE), which was extremely dangerous. So, we needed to keep an eye on him. When, after twenty-four hours, he was still in his room and vomiting, it was decided that for his safety, he needed to be flown back down to a lower elevation.

While I was bummed and down a crew member, my perspective on the highs and lows of the production were starting to shift. India is a very spiritual country, unlike any I'd experienced before. When in New Delhi, I had been awakened in the mornings by the Muslim call to prayer. Then, on a trip to the temple of the Monkey God, I visited a Hindu priest who christened me with a "Bindi"—a red dot representing the third eye. Now in the Himalayas, I was surrounded by Buddhist monks and influenced by their day-to-day coming to and fro. In comparison to the materialism of the West, India was more focused on the journey of the soul, and I was starting to better understand that everything I was experiencing was part of my journey. Rajat was a wise spiritual guide who believed in Karma and destiny—that is, Buddhist concepts related to a divine path and the idea that all actions and outcomes are connected to past choices and consequences. On multiple occasions, Rajat reminded me that "everything happens for a reason." And I was beginning to see his point of view. In Bible college, the idea that all things are predestined was something I would hear from the pulpit, but for the first time here, it was more than words; it was a lifestyle.

In resting upon my arrival, it began to sink in—I found myself letting go of my preconceived notions of how things "should go." There is more to this life than meets the eye. Releasing oneself from attachment to expectations and accepting reality as it happens was a big part of seeing beyond the mere physical experience. Taking in a deep breath and releasing it, I felt the freeing reality of this philosophy and started to move forward, striving to believe and have faith that it would all work out.

While my mind was opening up to new ways of thinking, it was not accepting the reality of the high altitude. Quietly alone in my room at night, I struggled in some nightmarish form between fiction and reality. In my dreams, I had visualized myself suffocating in my blankets, fighting to awaken, only to eventually wake up, gasping for air. Unaware of what was happening to my body, all I knew was that this experience was freakish, and I was so exhausted that I couldn't think straight about what to do. I would sit up and try to stay awake till morning. However, exhaustion would take over, and I would drift off again.

This process kept repeating. As my eyes closed and my brain shut down, I'd go into a state of hypopnea, the shallow breathing that comes with sleep. This naturally reduced my blood oxygen levels. And as the blood oxygen percentage dropped, it was then that my body would feel like it was suffocating, and I'd again enter the nightmarish dream state I mentioned. Snapping awake, I'd find myself gasping for air. To say the least, it was next-level torture.

As I struggled to survive the night, I was hoping that my exhausted body would acclimatize enough through the next day so I could carry on. Afraid that I could lose everything by being sent down to a lower altitude, I decided to keep the incident to myself. I was the leader of this production team, and if I had to leave, the project would be dead. It was the wrong choice—one that I would later find had life-long medical consequences—including pneumonia that would scar my right lung and an enlarged heart valve. Ignoring the dangers, I kept charging forward with reckless abandon. But by the second night, the nightmare returned, and I had a serious problem on my hands.

CHAPTER 11

Exhausted from having little to no sleep from the first night, I was not sure I would survive by the next morning. What had saved me before was my ability to awaken from my dreams, but on this night, I was concerned there would come a point when I would not wake up from it. If something happened while alone in a hotel room, no one would know till morning. So, feeling humbled and finally coming to terms with my destiny and accepting that I could not do this alone, I knocked on the medical team's hotel door and asked for help. Sitting me down, they placed a laundry clip-like device on my index finger to test my blood-oxygen levels. Getting a lower-than-average reading, they decided to put me on Diamox, a high-altitude medicine, and gave me a nebulizer to provide moisture and O2 for my lungs. Returning to my room, I laid back down and thankfully fell sound asleep.

After three days of resting, I was feeling nearly 100 percent, and it was time for me to start interviewing the runners. This would be my one opportunity to sit with them before they went into the next stage of acclimatization and training for the race. The training would start with driving the athletes to the trails where they would exit into beautiful high deserts to take a series of long walk-runs at higher altitudes as they prepared for the event. This meant that by the time they returned from both the travel and exercise, the runners would be exhausted and hard to get in for an interview. So, I needed to make these interviews happen in the window of about one day.

Visiting a classic Asian hotel down the street from mine, I reserved a large empty boardroom. I then purchased black cloth to block out all the light from the windows, allowing me to control the light from within. I was inspired by a documentary on the US war in Afghanistan called *Restrepo,* made by Tim Hetherington and Sebastian Junger. I wanted to give the interviews a format look of a subject lit with only a black backdrop and feel while creating a space where the runners felt safe and able to open up.

With the two tungsten ARRI lights that I had rented from a film school down in New Delhi and some work organizing the room, I was set up and ready to go with my list of questions in hand. Sitting down with the first subject, I was nervous but excited. I had done a lot of prep work to get to this point: overcame the hurdles of losing crew, suffered altitude sickness, and fought to go forward against all odds. With the microphone on and lights in place, I sat down with the camera rolling, then suddenly, without warning, the power went out, and the room went dark. Confused but figuring we'd blown a fuse, I tried to connect with the hotel staff to locate the fuse box.

After rushing through the courtyard to the front office to find someone who could speak English and help, I discovered that the city of Leh was not connected to the Indian national power grid. Being small, rural, and needing to conserve energy, they only maintained electricity on during the peak hours of the day in the morning and evening. I had not realized this because I didn't need lights during the day. Suddenly, and abruptly facing another challenge, it was testing my resolve.

Having canceled the interview, I felt knocked down again and began to feel like this entire project was cursed. With my crew in shambles, my limited resources of gear, and the interviews falling apart, I was laboring to accept my fate. So, I took a break in a small cafe for some tea and ramen noodles. As I recovered from my disappointment, I began to take stock of all that had gone wrong and considered that maybe it was happening for a reason. Maybe there was a bigger lesson to learn in all of this and a proper course that I was meant to be on. And if so, possibly all of this was happening to bring awareness to the path.

THE 2010 RACE AND THE STORY OF A LIFETIME

The evening arrived after I had spent some time mulling over the bumps and traversing the city by foot. And with it, the electricity was back on. Returning to my interview space, I had one final runner to interview, Molly Sheridan, who had been scheduled for the end of the day. She was the most unlikely of characters to be running a high-altitude marathon in the toughest environment in the world, and I wanted to know what she was doing up here. As I asked my first question, "Why are you here?" I was about to learn things I'd never known before and uncover one of the greatest stories of running that has ever happened in the sport.

One year prior, in 2010, Molly, like myself, had received a message on Facebook from Rajat. At that time, he'd been reaching out to hundreds of UltraRunners throughout the world, trying to build up responses and commitments to the race. Although hundreds of runners had said yes to attend, only Molly from the US and two other runners followed through on their commitment and actually showed up. The other two were Mark Cockbain from the United Kingdom and Bill Andrews, also from the United States. These three runners were the original pioneers of the first race—pioneers who believed in the impossible and took on a challenge that would nearly kill all three. However, before the lights came back on, it turns out there was much I didn't know about what these runners had gone through.

When I sat down on that day to interview Molly, I had no idea that she had attended the first race. Nor did I know that Bill, the second runner to sign up, was her lover and that they both had decided to perform a marriage ceremony mid-race, at the top of the high mountain pass Khardung La.

I also did not know that she had become critically dehydrated, collapsed on the side of the road, and was left unattended at a high altitude for a long period by her untrained crew. Fortunately, she was discovered just in time to be rushed to the hospital for an emergency IV drip in an effort to rehydrate her and save her life.

Furthermore, I did not know that Bill himself nearly died after running the first half of the race and was flown on an emergency flight out of Leh down to New Delhi for stabilization, then flown back to the United States for emergency gallbladder surgery to save his life. So, as the lights came back on, I was about to learn how much I had been in the dark with no clue about all that had happened on the ground during that first race.

When I began asking my questions and Molly began to share, it all became clear. As I delved deeper and she spelled it all out, all I could think of was how I would be able to retell this fantastic story about the pioneering and making of the toughest race on earth.

More about Molly before detailing her riveting account of what had happened—Molly was a tall blonde-haired woman with soft features and a gentle smile. Sitting across from me, I instantly liked her friendly laugh, which reminded me a little of Phoebe from the sitcom *Friends*. Easy to talk to and open to sharing whatever I asked, Molly told me how she got into the sport. At the age of forty, she attempted her first marathon, a distance of 26.2 miles, despite having had a prior knee injury. And her doctor told her that she was "too old to run that far." Always a fighter with a spirit to never give up, Molly was determined to prove him wrong, and she did. Against the doctor's orders, she returned to her training and finished her first marathon.

Inspired by that experience, Molly refused to accept being told no by anyone. That stubborn resilience led her to longer and longer distances and the sport of UltraRunning. Soon, she was completing her first one hundred-mile run of the Tahoe Rim Trail Endurance Runs before tackling much tougher routes like Badwater, which extends one hundred miles in extreme heat through Death Valley and Marathon Des Sables (MDS), a seven-day 150-mile stage race across the Sahara Desert.

Molly, never one to give up, reminded me a little of myself—a simple person from humble roots who'd been told no but refused to listen. When looking at her, you would never have assumed that she could accomplish what she had. When picturing iconic athletes for the sport of UltraRunning, I think of Dean Karnazes, David Goggins, or Ann Trason. However, Molly possessed something far greater. She was endowed with the spirit of pioneering, going where few have gone before, and the passion and desire to live out the dreams that would take her there. These qualities drove her to go forward through adversity, ignore the established opinions of the institutionalized, and dare to tackle great feats of running that might otherwise have been bypassed. I could relate.

As the interview proceeded, I was inspired by Molly's story, especially as she began to share more details of the mishaps and excitement from the 2010 race. One scene at a time, she recounted the fiction-like saga of that race, each segment worse than the last. I listened intently, intrigued as she admitted that

CHAPTER 11

the first error in her calculations was thinking that one could relax for a moment on this course. She and Bill, being in love, had planned for a simple ceremony at the peak of Khardung La. There at 18,300 feet, Bill, who had arrived over an hour before Molly and stood waiting longer than he should have for her to catch up, thus putting himself in danger. As she worked her way up, the dehydration began to creep up on her too. Finally meeting at the peak, the two spent a few special minutes together sharing vows. With crew cameras rolling, you could see the red scarf around Molly waving in a high-altitude breeze on a warm sunny day with what seemed like the entire world as a backdrop, as seen from the mountaintop. Finishing their vows, they said their goodbyes and started back down on foot toward the other side. What nobody understood at this point was that they were both in danger from being exposed too long at high altitudes and were doomed to suffer the consequences of pushing their bodies beyond their limits.

Some forty minutes later, and just a few thousand feet down from the top, Molly was still at high altitude and alone on the hillside, her crew having been ordered to go ahead and wait for an indefinite period. Being untrained in the sport, they drove too far into the distance. And Molly, hit by dehydration, couldn't make the rendezvous point, so she collapsed in the dirt on the side of the road. After she lay there for nearly an hour, the car finally circled back. When they discovered her, she was pale, weak, had little energy left, and was struggling just to stand. So, the crew helped her up and into the vehicle and then raced down the mountain to get her to a hospital for an IV, hoping to see her recover.

Meanwhile, some thirty miles farther in the distance and an hour ahead on foot, Bill had been pushing hard but suddenly was overcome with pain and now faced his own problems. Curled over in a ball, he was found leaning on a roadside railing, grasping for dear life. With the crew recognizing the danger, he was the second to be rushed to a nearby hospital where his pain only seemed to get worse.

At this point, Mark was the only runner left standing on the course, but he, too, had faced a series of trials. While summiting Khardung La in front of both Bill and Molly, he had survived a small avalanche that had slid across the high-altitude road, pushing rocks over the edge of the road and plummeting deep into the distance. Unwilling to stop his trajectory and desiring to get off the peak as soon as possible, Mark charged over the snow and continued far ahead, ignoring the military personnel who were stopping cars from traversing the route. After having worked the better part of a day through the Leh Valley and into the night, he had made it some eighty miles and was beginning his climb up to Tanglang La at 17,300 feet, seemingly on course to make it to the finish.

In the madness of Molly getting an IV, she sat recovering in the open emergency room as an Indian tourist who'd flown up from sea level that day died on a gurney. To her recalling it, it still seemed fictional. On the course from the start was a pseudo-journalist named Robert Wier, who had been recruited by Molly. He was there reporting and functioning as crew. In expereienced in the sport and high altitude, he went into full panic mode. For him, this sport was foreign; what these people were doing seemed insane, and he was collapsing under the chaos perhaps because he was also suffering from high-altitude exhaustion. Weir, seeing Rajat at the hospital, exploded in a desperate rage and began yelling, "STOP THE RACE!" Face-to-face with the man, Rajat, a doctor and UltraRunner, having grown up going to school in the foothills of the Himalayas, remained ever so calm, weathered the outburst, and refused to counter the runner's decision to remain in the race. Instead, he kicked Weir off the course and out of the event.

In the chaos of the battle, Bill's condition worsened. He was now white as a ghost and clutching for dear life. Grabbing hold of the ear of a crew member sitting by his bedside, he repeated over and over again, "I'm going to die. I'm going to die."

The hospitals in this remote part of the world are limited, so it was decided he needed to be flown to New Delhi to stabilize and be diagnosed. Meanwhile, to my surprise, Molly, who was

CHAPTER 11

kept unaware of Bill's condition to this point, had re-entered the race, having recovered post-IV and progressed some sixty miles into the run across the Leh Valley. There in the desert, Rajat found her and pulled up in a support car to give Molly the news about Bill. Feeling compromised between leaving the course and seeing Bill off to New Delhi, she decided to drop out of the race, realizing that she couldn't imagine continuing knowing that he might die. While the race was then over for them both, the world seemed to spin upside down for the final runner, Mark, as he reached critical exhaustion.

For Mark, the night had arrived after covering nearly one hundred miles of running. And just after he'd reached the summit of Tanglang La, he desperately needed to rest and warm up. With no physical place to go, he decided to climb into the back of the crew's SUV and lay down. Balling up in the boot of the vehicle, he closed his eyes and quickly fell into a slumber as the car continued to run to keep him warm. The danger was that at high altitudes, the fumes do not easily dissipate, and unbeknownst to him, he was starting to asphyxiate—a condition wherein the body is deprived of oxygen and can lead to unconsciousness or death. While dreaming, he caught a whiff of the fumes and snapped out of his exhaustion. Realizing the danger and feeling the general lack of oxygen at high altitude, he kicked open the door, stumbled onto his worn feet to gulp some fresh air, and started the long, faithful process of dragging his near corpse the final twenty miles to the finish.

Back in the interview room, hearing Molly weave her story, I had listened with bated breath, fully engrossed in the details. In the previous hour, Molly had elaborated a play-by-play story about which I'd heard nothing prior to this moment. Rajat had kept the details close to his chest, and never once, on any Skype call, informed me about what had really happened. I had only heard about Mark's finish. Now hearing this, I was amazed and needed more details from the race director himself.

After completing the interview, I quickly ran to find Rajat to ask more questions. I knew that he'd invited a few journalists to the 2010 run, and I was curious to know what footage they all had captured. I now wanted to see firsthand everything that I had just been told. Asking him directly, he finally began to open up and, with it, revealed the archive of all that had been caught on tape.

From my perspective, the experience sounded terrifying, but for him, it seemed like just another day in the park. He had been there, taken both Bill and Molly to the hospital, and escorted them both to the plane. For him, everyone had survived, and the race was done. And while death seemed to have been around every corner, none of it seemed to have phased him. He had continued to say over and over again that "anything was possible," and he lived what he believed. I realized that death is always a risk when going beyond the known and pushing into the unknown, and this was a whole 'nother level.

With the details of the pioneering run in my head and discovering the large deposit of archive footage that Rajat had collected, I knew what I had to do. I had to tell this story. Now, all my struggles with filming the 2011 race seemed to be pointing at this discovery, and I was beginning to think, like Rajat, that "everything happens for a reason." This was a story of a lifetime and one that I had to tell. To do so, I would need to navigate the impossible. I would need to schedule interviews with runners, journalists, and crew and shoot the visual recreation of portions of the course to fill in the blanks where the archive video could not. I would also need Mark, Bill, and Molly to travel back to India and the farther reaches of the world, and to do all of this would require funding from people other than myself. Above all, I would need to entertain doing the impossible and believe that if this was my destiny and my divine intuition was real, this was meant to be, then go forward boldly, trusting that all pieces would fall into place.

Barry capturing Rajat in 2010 in Leh Valley, India

Ellis Island, New York (2012) –
Barry and Stefania after moving from Italy.

"Ask and it will be given unto you; seek and you shall find; knock and the door will be opened . . ."

—Matthew 7:7 (KJV)

CHAPTER 12

MODERN MIRACLES AND AMERICA (2011 - 2012)

In elementary school, I specifically remember the words of Orville Merillat, a very significant figure from our town. In 1946, after serving in World War II, Orville and his wife Ruth started a small woodworking shop in their garage in Adrian, Michigan. This modest beginning grew into Merillat Industries, which became the world's largest cabinet manufacturer with numerous plants across the United States. Having amassed wealth beyond his wildest dreams, he used his fortune to support multiple Christian organizations, including Focus on the Family, World Vision, and Youth for Christ. But the organization he'd finished that impacted me the most was the private Christian school that I had grown up attending known as Trenton Hills (later to become known as Lenawee Christian). As a third grader and famed "class clown," I won the heart of his granddaughter Trisha Merillat, with whom I got caught holding hands while on a school bus field trip. I remember the childhood embarrassment of it to this day. When, on one occasion, Orville visited the school to speak, I remember him saying to the students, "There is no dream too big for God, so dream big!" For some reason, in all the things I heard said over those early years, those words stuck and rolled over and over in my mind. As a result, I have been dreaming big ever since.

By this point in life, I'd gotten good at dreaming big and was getting more confident at living out my dreams every day. Each time I jumped off a cliff in life toward living a big dream, I'd learn something new from basic trial and error. In the process, I formed a strong hypothesis about how dreams lived can be executed and realized a practical breakdown of how to map your course. In reading Joseph Campbell's book, *The Hero with a Thousand Faces,* I learned about a method used for writing stories of myth and legend and discovered one of the best breakdowns for living out one's dream and casting yourself as the hero. The hero's journey, as it is called, is configured with five classical phases of storytelling that align with the process of executing any dream.

The five phases are as follows:

1. The Call to Action;

2. Departure from the Normal World/Crossing over the Threshold into the Unknown;

3. Trials and Temptations;

4. Crossing over the Abyss; and

5. Returning Home Changed from Your Passage.

The value of understanding this circular journey is that it parallels much of life's experience. It can help you map where you are on your course to living a dream, offer insight into where you going, and assist in self-correcting in your aim to live out your dreams.

CHAPTER 12

For example, in my marriage to Stefania, I had completed the full hero's journey as it relates to falling in love and the physical ceremony. I had gotten the "call to action" on the beach in Costa Rica; I had "departed the normal world" by leaving my home; I had "crossed over the threshold" by selling all my possessions and moving to Italy; I had faced "trials and temptations" in the language, the culture, my work visa, and in many other unmentioned incidents; and I then "crossed over the abyss," the major turning point and test of the "hero"—which, for me, was my marriage proposal and choosing between going forward and facing my fears or turning back. Ultimately, after our wedding and with plans to go back to America, I was "returning home changed" in a most literal and metaphorical sense.

THE RETURN HOME AND THE HIGH

While the previous three years in Italy were an amazing experience, I had desperately wanted to return to a career and knew that to do so I needed to go back to America. In the months and years that had passed since moving to Italy, my relationship with Anna, Angelo, and the family had grown. In many ways, I had found something that I'd always wanted in their family, there love and acceptance felt much like the childhood fish fries on Grandpa's farm from just being together with the family. In my years living in Italy, Anna taught me how to make Limoncello, and Angelo took me to make the family olive oil, and together we spent weekends visiting siblings and friends around the countryside. It was something I deeply treasured. However, Stefania and I were still struggling to make ends meet day-to-day. Between us, we could barely cover the cost of rent and food. On any given month, I was lucky to have some loose change to spend on entertainment for Stefania or just to buy a beer for myself at the local cafe. Angelo, hearing of our struggles, worked to give me options to make more money. He spoke with Anna and Stefania about having me take over one of his local restaurants or cafés. He also offered to help us purchase our first home. While this was extremely generous, I knew that this would mean an end to my career in media and filmmaking and a deeper commitment to the family and Angelo. I still had a strong desire to create films, work to build my own fortune, and fulfill what was known as "the American Dream." While it was tempting to accept the offer and stay, I knew that if I accepted it, we'd never get out of Italy.

In our desire to move, there was one big challenge, money. When I first moved to Italy, I had very little cash and spent most of what I made filming the production in India. So, while the idea of going back together to the States sounded great, the reality seemed almost impossible.

As if my hopes to return to America and the challenges of funding the move weren't insane enough, I also needed to continue forward on my vision to complete telling the story of *The High: Making the Toughest Race on Earth,* which had to happen in just over ten months now. This felt doubly impossible to do. The biggest reason was that I needed to travel to all the key characters' homes and conduct interviews. These characters were spread out around the world: Rajat Chauhan, the race director, was in India; Bill Andrews and Molly Sheridan, the runners, were in Nevada, USA; Mark Cockbain, the solo finisher, was in the United Kingdom; Ben Arnoldy, a journalist from *The Christian Science Monitor*, lived in Boston, Massachusetts; and Robert Wier, the quirky independent journalist and crew support for Molly, was residing in Michigan. Getting to each of these people would cost a small fortune. I also needed to return to India and capture recreations of the key scenes of Molly getting sick, Bill getting rushed to the hospital, Mark nearly being asphyxiated at 17,000 feet, and the general beauty and course footage that would make the film great. Between America and India, my mind worked full-time, and we were very much in the space of doing the impossible.

To do all this right would require the budget of a feature film, which I didn't have. What I did have was a belief that if I followed the basic steps of courageously taking on my dreams and mapping out my course within the hero's journey, I could turn any dream into reality, including this one. Moving one foot in front of another, an idea popped into my head. On the cusp of the age of crowdfunding, I had watched, time and time again, people with good ideas raise funds from friends and followers who wanted to see them succeed. Reviewing the *Kickstarter* website, I decided to try my hand at raising my own money. Mixing the footage from the 2011 race with the interviews I'd shot of athletes, I strove to articulate on camera the cause, then edit an explainer video together with a teaser of the film and finish with a call to action. With the computer uploading the piece to the platform, I punched away at the keys, squinting to be sure I had all the correct details in place. Cutting and pasting the full scale of my email contact list that I'd collected since college, I blasted the project out to over 500 people and aimed our goal of raising at $5,000. If we got it, it'd be enough to cover the airfare and travel to India and then some.

THE CEREMONY ON THE MEDITERRANEAN

Having rushed to get married for legal reasons and permission to work, our first wedding, as it has come to be known, was in a state building with a few close family and friends. My American family had only heard details about this wedding and seen photos but did not experience it. Talking to Stefania, we wanted a celebration ceremony or even just a party where everyone would have the time to come and participate with us. Hearing this, Anna and Angelo, being ever generous, got behind the idea of the event and helped us with funding, and Stefania went to work.

After weeks of hard work by Stefania and friends, in August 2011, we stood joined together at sunset on the Mediterranean Sea. With a light breeze blowing and my pulse racing, we celebrated the event with our family and some of our closest friends. I danced to both American and Italian music with my feet twisting the beach sand onto the patio. Having worked diligently on speaking the language, I translated for my parents, helping them connect with all the Italian family. The embrace and love of that community was rare, and all our hard work to get here had paid off since our meeting in Costa Rica. We had been tested, and our love proved strong. We'd overcome a great physical distance from the US to Italy, and we'd bridged a culture gap and tackled a language barrier to form a rare bond that was stronger than anything I'd ever known to this day.

After recovering from an amazing evening, we were adding up all the gifts and money people had given us. For the past two years, we'd talked about moving to America, where I could return to work and Stefania could start to live the American dream. However, to this point, it was not realistic because we did not possess the funds to go. Sitting around the table that night with the booze and champagne still coursing through our veins, I flexed the cash through my fingers and added it all up. By the time the night wrapped up and we'd gone to bed, we'd counted enough to make the passage back, buy a car, and get an apartment. For Stefania, it was a dream come true; however, for me, it was bittersweet. I would be leaving a place I'd come to love and a sense of family I'd deeply desired to have. In exchange, my career would again see the light of day, and my dream of making a film of notoriety might actually come true.

GOODBYE, OLD COUNTRY, HELLO, NEW WORLD

Immigrants coming to America has been a romantic notion that our country was founded upon. From the Ken Burns documentary *The Statue of Liberty*, I could mentally see the black-and-white film capturing Ellis Island and the lines of people waiting to pass through inspection while the American flag waved high overhead. Growing up a naturalized citizen, I had an endearing love for our country's history, and the stories of immigration were a big part of that in my family. Going back only a short time, you could find that my great-grandma and great-grandpa from both sides of the family had moved to the States from various parts of Europe. To do so, they had given up their homes, extended family, and future for dreams of a new life. Now for the first time in my life, I felt what my ancestors felt, and it was a great honor to be able to share that common experience with Stefania.

Preparing, planning, and working through every detail, I decided that this dream would have forethought. Being married, I wanted to be responsible and avoid jumping off the metaphorical cliff as I had so many times before unless absolutely necessary. So, reviewing the map and speaking with friends, we decided the best place to move and settle down would be New York City. While this was a dream of Stefania's come true, I had friends there, and work was plentiful in the Big Apple.

Reviewing our finances, I knew exactly what I had to spend and what I needed to spend it on. From day one, the plan was to purchase a flight, find an apartment, buy a car, and then start looking for a job. Counting our dollars, I knew we would be cutting it close. In my mind, after all was said and done, we had about two to three months of income to live on before we'd run out. Between departing Italy and getting established in New York, there were a lot of unknowns, the biggest of which was employment. I had not held a professional media job for nearly three years. While I'd kept myself busy on personal projects and mostly volunteered on sets, there wasn't the physical experience of working with people in the business. Nor was there the lingo, jargon, and trends that come with being around it. To make it in New York, I'd need some divine intervention and the faith that things were going to work out somehow.

For me, faith is a highly misunderstood and extremely underrated tool. In modern times, some see it as an extension of the church to dope the masses. For many atheists or science-based people, it is believed to be a crutch for the weak to avoid dealing with the realities of life or, more importantly, death. However, I believe that faith is as essential to life as the heart is to the human body and manifests in all humans every day of their lives, often without their acknowledgment. Basic common uses of faith that are dismissed as not applicable include things like faith in people showing up on time, faith in our systems of government to protect us, or faith in the truths of science and medicine. Daily, humans traverse from the known to the unknown in the smallest of ways, and without faith, they'd never make the passage.

On a larger scale, within the lives and hearts of monotheists, polytheists, and atheists alike, there is arguably an underlying faith that what they believe is true. No religion or belief is without that which is known and that which is unknown. Some things will always exist that just cannot be physically proven. At the furthest reach of any belief is where faith begins. Painfully, I feel that we live in an age where few will concede what their faith is in and own up to what they believe. A void in any system of belief is doomed for trouble.

THE FIRST MIRACLE OF NEW YORK (2012)

Landing in New York City was a wake-up call. The streets of that town are hard, and the people are harder. The black tops were full of yellow checkered taxis racing past endless rows of skyscrapers—like the ones on Fifth Avenue, born in the early 1920s, feel like they are shouting out, "We're in a boom America. Join us; it's the Roaring Twenties!"

In that space, you feel like you could see Babe Ruth driving by in a Packard 426 Roadster, the kind with the long nose that you'd see in a mobster film. Everywhere you walk from that period and beyond carries with it the strong voice of Frank Sinatra that echoes out, "If you can make it here, you can make it anywhere"—a line that is as compelling and intimidating as the great city itself.

Flying back to the USA, I knew it was "go time." First, I had to pick up a car that we'd purchased in Michigan from my brother-in-law. I then drove to New York ahead of Stefania to get an apartment before she arrived. Out in Long Beach, a small island near Queens, I crashed on the couch of my friend Chris, who I'd known from the LA era. He had once been my roommate and wingman on the West Coast, but things were different now. He had to juggle between work responsibilities and a family. Being greeted as "Uncle Barry" was a blessing to me, but for Chris and

CHAPTER 12

his family, I was an added guest requiring a shift in routines and work. So, I knew I needed to make the visit short and find an apartment fast.

After hiring an agent, I hit the streets—from rubbish to riches, I looked at it all. Walking in and out of the doorways, yards, kitchens, and bathrooms of twenty or more units in the span of a week, I ended up finding a place in a brick building with marble floors and a green door just a block from the beach— meaning I could finally return to surfing and Stefania to get some sun. Finalizing the deal, I counted out the cash for our first and last month's and a deposit to the landlord. I could feel my bottom clinch as I handed over the $8,000 down payment in exchange for the key. Now, the clock was ticking, and it was time to get a job.

Every day, I woke up and searched the Internet want-ads, looking for a job in production. With eyes glazed over from staring at a computer screen, I manually filled out online application after application, clicking and sending over and over, hoping for the best. My biggest concern was my patchy résumé. Even though it included work I had done, it had gaps. For example, it revealed in work in film in LA, then a gap for my time teaching in Santa Cruz; in San Francisco, work in television, and then a gap for my time in Italy; in Italy, work at the famed Cinecittà Studios for a day as first AD on a large film funded by the Romanian mafia, and work completing my own film, having taught myself to edit using the Final Cut software program.

I built up all of this in my résumé as experience, but apples for apples, it had its gaps. All in all, my work history was not ideal for any HR department, especially those in Manhattan, but I continued to fill in the forms and send the applications, praying that things would work out.

Looking for work in the cement jungle of downtown New York, I plodded along, shoes dragging on the sidewalk as people pushed passed on my right and left. In the distance was a silver reflection beaming off the peak of the Chrysler Building as I turned to enter the iconic Bryant Park on 14th Street. There, I found some refuge next to the granite base of a fountain cascading with the sound of water on water.

I was feeling overwhelmed and in doubt about everything I believed about dreams and reality. I'd only had one interview in the span of a month, and that interview felt like a long shot at best. The interview was at the National Hockey League (NHL) headquarters in downtown New York City for a position as an editor on their daily sports show *NHL Live*. I clearly remember that giant iconic silver NHL emblem rotating in the entryway. Sitting in the lobby, I was one of many people sauntering in and out of the interview room. Upon walking into a giant boardroom, I looked down at the table, which seemed to extend for a hundred yards. In front of me, there was a stack of résumés nearly an inch thick, with mine sitting on top. I felt overwhelmed feeling the little self-confidence I had shrivel up inside me as the words of one of the three men at the table started to roll off his tongue.

To this point, I had only worked as an amateur editor on my own projects. Being an anti-establishment type, I had taught myself how to edit. The challenge in the interview was that I wasn't sure what was relevant to this position. I had never been a show editor on a major national television program, let alone an editor in general for a company. And I did not know what was required to get the job done. However, what I did know was that I desperately needed work, and this was a do-or-die situation. So, I decided to adopt the motto "say yes and figure it out later."

Finishing the interview, I walked out of the giant forty-fourth-floor boardroom and exited the building out onto the street. Turning to look up, I was overwhelmed by its size. This structure was a giant in a land of giants. Stepping close, I placed a hand on the massive steel frame that reached into the sky and said a prayer. There was little chance that I would get the job; I was underqualified, but if there were a chance, it would have to be divine, at least in part. With that, I headed home to Long Beach onboard a Long Island Rail Road (LIRR). Walking in the door, I could smell the fresh pasta sauce and see my ever-positive wife shining with pride. "I am sure you got the job," she said while setting in front of me a delicious bowl of pasta. A wife who supports her husband through thick and thin, what a treasure.

INDIA AND PREPARING TO RETURN TO THE HIGH (2012)

In the evenings, when I wasn't looking for work, I was also working to connect with the other runners, crew, and participants who were part of the first edition of *The High Race* in India. Talking to these people via Skype inspired me. They had all been pioneers in a sense. They had each believed in something that few others did and set out to accomplish it against all odds. For me, they represented exactly what I believed in—accomplishing your dreams—and they had set out to accomplish that in their own ways.

Having connected with everyone who'd run the 2010 race, I worked to make plans to return to India and schedule interviews to hear their stories. I needed to properly allocate the funds from the Kickstarter Crowdfunding campaign that I'd launched before leaving Italy. After blasting out word of the fundraiser, I waited for the two-week window for people to donate. When it finally did, I had raised 6,371 dollars contributed by eighty-seven people. I was blown away by their support and generosity. Not only had I reached my goal, but I had also surpassed it!

After finishing the fundraising and prior to leaving Italy, I used a portion of the money to travel to the United Kingdom and interview Mark Cockbain. He was a short hop from Rome, and we spent a day together interviewing and learning more about his run. Now reviewing every inch of the archive footage that I'd acquired from Rajat, I made note of each runner's attire, state of mind, and level of exhaustion as the race progressed. This footage would help me build out the scenes, structure my questions, and uncover the key moments in the film needed to capture the audience.

In my research and plans for interviews, I called Bill to speak for the first time. During the call, he shared that both Molly and he were planning to return to India to run in the 2012 race. Bill,

THE UNKNOWN ADVENTURER

like Molly, was determined to finish what he'd started in 2010, and Molly decided she would travel with him for moral support. This was fateful serendipity at a level I could not explain. I would be able to interview Bill, Molly, and Rajat and shoot recreations of all the major events in one location, Leh, India. It felt miraculous. I could not believe what I was hearing.

As exciting as the new developments with the film were, I was still terrified by the reality that it may never happen. Unemployed and paying $2,500 in cash a month for rent, my funds were nearly gone. I had been on the job hunt for six weeks and still had nothing. While I aimed to be responsible with the funds I'd raise from Kickstarter, part of me knew that if push came to shove, I'd rob Peter to pay Paul before going bankrupt in New York.

Peering out through the thin white window shades of my apartment in Long Beach, I would watch as cars of busy people traveling to work and home run back and forth along Broadway. With a thin haze between myself and the world, I began to think of a plan B in case things didn't pan out with finding work and living in New York. Mulling over and over in my head what options I had, I knew that the reality was that there wasn't a plan B—worst case—we could live in my parents' basement. Doing so would be a painful setback. I imagined my family would look down on me, and to my supporters, I'd be seen as a scammer taking funds for a project that didn't exist—an idea that scared me deep into my core.

Alternately, I thought we could go back to Italy. There, Angelo and the family would pick us up and get us on solid ground, but if I went back a second time, I knew I would never get out. And I wasn't prepared for that.

Pacing back and forth, searching for answers, I was so deep in thought that I didn't hear the phone ringing. Yelling out from the kitchen, Stefania alerted me, and I snapped back to reality, just in time to pick it up. Answering, I said, "Hello."

On the other end was a man named Russ who jumped immediately into conversation as if we'd had an ongoing dialogue for weeks. His tone and comfort caught me off guard as he said,

CHAPTER 12

"Barry, I have been trying to get you in here to work, but HR is taking forever with the paperwork."

Pausing, I was still trying to figure out who exactly Russ was when he went on,

"Let me ask, can you start work on Thursday?"

Then, it clicked. It was Russ, the director of operations who'd interviewed me at the giant table in the NHL headquarters downtown. In a flash, I wondered with the comfort and tone of his voice, did he call the wrong applicant? Pausing, I nearly asked until I realized that I didn't want to alert him to his error. If I was correct, it was better just to take the job.

Pulling myself together, I responded, "Yes."

"Good," he went on, then started briefing me on pay and hours.

Hanging up, I let out a sigh of relief, not knowing whether to laugh or cry. I was almost bankrupt, and in one phone call, everything had changed. Years later, after looking back on this moment, I called it the first miracle of New York. For now, I had a job, and my mind instantly shifted to figuring out how to go back to India in under six months.

THE NHL ICE AND THE INDIA HEAT

From day one, I jumped headfirst into working at the NHL. I woke up early, was prepared before the team was in the office, and pushed myself hard to grow with the goal of being one of the best editors on the show. Every day, I woke at 6:00 a.m., rode the LIRR train to Penn Station, and walked to 44th Street and the Avenue of the Americas. Passing security, I would join a crowd of people chasing up the elevator, exiting on the seventh floor. After some prep, I would hit the morning meeting at 10 a.m., get my rundown of assignments, and go to editing, chasing the 5 a.m. deadline when the show went live. Over the next eight hours, I would edit a series of show packages—from the news to highlights and reels, generic player b-roll to the most coveted assignments—the show open. The show open is the sizzle that kicks off at the top of the hour and is assigned to the best editor—a title that I wanted to earn.

At 5:00 p.m. every day, no matter whether your work was done or not, the show would go live. Instantly, you would have producers yelling in your ear for their segments. We were doing everything we could to be sure they were uploaded and on the server for them to pull down at each time slot of the day. Most of the time, everything got done, but sometimes it didn't. If it was your piece that got dropped at the end of the day, there was hell to pay from the producers for that mistake.

By 7:00 p.m., each day the show would be wrapped. I would pack up my stuff, walk back to Penn Station, and return home the way I came, just to wake up the next day and do it all over again. It was hard work, but I was cutting my teeth as an editor and making it happen in the Big Apple.

With so much to do, the time flew by. Days became weeks, weeks became months, and soon we were coming up on August, which meant the second running of *The High Race* in India was near. I was trying to figure out how I could get there without losing my NHL job, which I loved. It had been very hard to get a job in this city, and leaving for three weeks would most likely not be possible. In fact, I would either get fired or have to quit, and I couldn't justify doing either.

Ironically, however, I would not have to do that. Thinking back, I remember that on the day I was hired, the previous editor who was training me shared one final note as he packed up to leave at the end of his tenure. Standing at the edit bay door, I was now in the edit seat, and he was ready to leave. Looking at him in the eye, I could tell he was bummed to go. And he shared with me something I wished would not come true. He said, "Do yourself a favor and don't get too comfortable because no matter how good you are, they are going to let you go at the end of your contract."

Surprised, I watched as he wished me well and left. As the NHL season came to an end and my contract was closing out, I, too got my pink slip. No matter how much I loved the job, how hard I worked, or how much people loved me, to the NHL corporation, I was just a body, and by the rules set by national legislation, they would be obligated to provide healthcare for that body after six months of employment to save money—they just turned and burned editors out. That's what corporate America and big government policy is all about. So, at the end of my sixth month, with the swiftness of death and the emotion of a machine, they sent me packing and brought in someone to replace me.

Once again, just a few weeks before the trip to India, I was out of work. Free to go, I booked my ticket and hired a young director of photography named James to come with me. I was completing my obligations to myself and those who'd funded the project on Kickstarter. Much like the first miracle of New York, I believed that the timing of Bill and Molly returning to India for the 2012 race and the perfect sequence of releasing me from work obligations to go was the second miracle of New York. God, in His mysterious ways, was at work.

RECREATING THE ORIGINAL STORY IN 2012

On my return to India, I would now focus on telling the story of the original running of the race that happened in 2010 before I arrived in 2011. This time, however, I was much better prepared for what I was getting myself into. I knew the altitude. I knew the power issues. I knew the equipment I needed. And I knew the level of exhaustion I would be feeling. In short, I was ready for what this brutal part of the world was prepared to dish out.

Flying into New Delhi, Rajat provided us with a room in his house, and we spent the first few days filming the first recreations of the film. The recreation would be composed of Rajat being inspired by a map to do the run, then his reaching out on Facebook to runners throughout the globe, followed by a series of shots of him speaking in person and on the phone with the Indian military, asking for permission to do the run. It was a blast working with his relatives and family to shoot a few of the scenes. I connected with the culture in ways I'd never expected to do. One night after a long day of work, Rajat invited us to

THE UNKNOWN ADVENTURER

his dining room, and we sat down at the table with his wife and kids to eat.

Entering, I got the delicious smell of Indian food that filled the room, and the kids were placed around the table in grand wooden chairs. I found my place, was served some rice and curry, and then began to eat. The experience of eating with this Indian family, which I'd come to know and love, was intimate and the conversation fascinating. I grew up in a Christian home with Christian iconography and visuals placed around the house. This was a Hindu home with Hindu iconography and visuals. In entering this portion of the home, I'd passed a small altar where Rajat's wife placed a daily offering to the gods. As fascinating as it was, it was foreign to me, yet there was something very common for us all to share.

While eating some amazing food and talking with the parents and kids, I asked Rajat's wife what things were important to her on a day-to-day basis. Interestingly, she listed what every wife in any home in America would say. She worried about the kids growing up and getting a good education and the safety of her family. Then, in an entertaining way, Rajat shared that his wife liked to spend too much money shopping—a common struggle in every relationship. The fact that they shared the same kind of struggles I found in my own childhood home was eye-opening to me. Here I was, on the other side of the world, a very different world than mine, and yet the human struggle was very much the same.

After a few days in town, the runners for the 2012 race started to arrive, including Bill and Molly, for the third running of the race, which was happening sequentially with our shoot. After the excitement of getting to meet Bill for the first time in person, we traveled to an "ad-hoc" hospital room that looked like where Bill and Molly were both hospitalized. There, using some basic hospital supplies we'd acquired as props, we recreated the scene that Bill felt was eerily similar to the bed he had nearly died on. It was hot, and everyone was still suffering from jet lag, but I was loving every minute of seeing the vision come to life.

CHAPTER 12

After getting Bill into position, we started to roll the camera, and for the first time in my life, I felt like a director on the set of my own film. Finally, I was living a dream that I had envisioned when on sets in Hollywood some twelve years earlier. Seeing the vision come to life was a surreal feeling that transcended time and space. All I had done and overcome to get here was hitting me at the same time. I was flooded with emotion and excitement and doing my best to both embrace it all and get the job done.

After two weeks of working in the blazing heat near sea level in New Delhi, we boarded a plane to the magical city of Leh on the way back up to the elevation of 11,000 feet. Surrounded by mountain peaks, I stepped off the plane, walked out onto the airport tarmac, and took in the thin air and beauty of the place. It felt great to officially be back again.

In 2011, I had interviewed Molly. Shortly after, I traveled to England to interview Mark, the solo finisher of that original race in 2010. However, I had not yet interviewed Bill, the time of which was now at hand, and I was very excited to hear his story. I knocked on the small wooden door of his hotel room, and he, tall and lean, exited from the darkness within it. We walked together to the space that I had prepared and placed a mic on him for audio. Then, he sat down, and we started rolling.

To add to the mystique of his UltraRunning feats, I also discovered that Bill was a scientist. He had spent his life's work studying aging and working to slow down or even reverse the process in human beings. Crazy as it sounded, his motto was to "cure aging or die trying." I thought it was pretty funny and a twist of irony since he'd almost died on his last visit here and was back to face death again. Listening to him passionately speak on the topic, I was even more amazed to discover that this man actually believed he could reverse the "disease of aging" as he called it. And he had made great progress in the field, both in understanding it and providing solutions. It all made perfect sense that he had attempted this impossible race and was so obsessed that he returned to take it on again because this man had no concept of the impossible. He only knew "Try, try again."

THE UNKNOWN ADVENTURER

In the ten days of acclimatization before the upcoming race, I needed to film the recreation shots. So, I'd build a storyboard of the exact shots I wanted to do. I also wanted it to feel as authentic as possible, so I got everyone in original gear matching scene-for-scene from the archives. And I worked to capture the runners in a natural environment to avoid any need for acting. Driving up to high altitude, we spent the day working the route up to and over Khardung La. The air was thin, our heads felt like they were inside a helium balloon, and everyone moved slowly as we worked. Having Molly lay roadside, we worked to capture the look and feel of her collapsed from dehydration. Rolling herself into the dirt, she willingly played the part. Then, we needed Bill to roll over the side of a guard rail, where his crew had discovered him collapsed before being raced off to hospital. Capturing the beautiful landscape in the background, James, the director of photography, framed each shot as the first day of the recreations went off without a hitch or any signs of altitude sickness.

I hugged Bill and wished him well in his upcoming race and finished one last scene with Molly near the town of Rumpse, where she had finally returned to find Bill. Closing in on the last two days of shooting, we turned our focus on the final scenes and the remaining runner, Mark. In considering Mark's scenes, I had first tried to convince him to return when I visited him in the United Kingdom. Having been traumatized by the near-death experience, he had no desire to return. At a loss, I needed an actor who looked like him. Fortunately, I came pretty close to looking like him from the side or back. So, I cast myself in that role. Reviewing the footage, I noted that Mark had worn a hat to protect him from the sun for much of the race. This was perfect and allowed me to hide my face from the camera. Preparing for the roll, I cut back on my pasta intake to drop some weight and asked Mark if he'd send me the gear he had run in. He kindly did, so I dressed identically like he had looked on the course, right down to socks and shoes. I even rented an identical support vehicle that his team used for the race so that not a thing would be missed.

CHAPTER 12

Ahead of me in the distance was a sharp mountain peak towering above a squiggly line of switchback roads that worked their way up to the top. Leaving behind the small farm homes of Leh Valley, we started to work our way up to the harsh peak of Tanglang La. It was here that Mark had been coming undone, not staying firm on the course, and Rajat had guided him toward the peak. As night set in, we captured the last scenes at 17,800 feet, where Mark had passed over the top.

With the crew exhausted, we wrapped up for the day and drove down to the Nuru Valley to spend the night at 14,000 feet in a gypsy tent made for truckers driving up from Manali. After eating some ramen noodles and warming up to some butter tea, we fell asleep on two-inch mats and laid in circles on the floor in a giant "circus-style" tent colored all in white. All around us through the night, truckers parked, entered, and fell asleep. Isolated from the world outside, I felt more like a local hero than ever. Something about the raw suffering and intimacy of the dormitory sleeping broke down any barrier that I'd once had, and I felt at home.

The next morning, the sun cracked over the high desert, leaving landscapes that felt like I could see into eternity. It was early, and we were all exhausted and moving slowly, but we needed to catch the sunrise for Mark's scene of crossing the finish line. Packing up our truck, I drove down the road until Rajat said to stop. The end of the race was in the middle of absolutely nowhere on earth. Rajat literally could not have had the finish line in a more obscure and barren part of the world. But it seemed fitting for the race as there had been an ongoing theme of suffering for only personal glory. No selfies were happening here; these people did it for the love.

Camera rolling, my DP James called "Action," and I, playing as Mark, pushed myself forward toward the end of a miserable 133-mile two-day run. In character, I imagined the pain in my feet, my joints swollen, and my eyes half open as I dragged myself alone, one step at a time. The finish here was like a finish in life. In the last thirty seconds from the race archives, I watched Mark struggle to arrive at the finish. He reminded me of Rocky at the end of a long fight, just trying to go the distance. As I crossed the line recreating the scene on camera, I, too, felt like I was finishing an insane race. There in the desert, with nothing around, I had finished the recreations of the film. As I heard James call out "Cut," I now could rest for a second and reflect on how far we'd come.

Concluding the shoot, I spent the final few days enjoying the region and cheering on Bill. Waking up early one morning, I hiked the 200-plus steps up to the Buddhist temple near Shanti Stupah and meditated to the gong of Buddhist monks for over an hour. In that place, meditating, I achieved a moment where time and space seemed to transcend all things, and I was fully present.

In my mind's eye, I began to travel and could see the full passage of my own hero's journey to this point. From childhood to the present, I could see the timeline of the young boy with a big imagination struggling in the classroom as he daydreamed out the window, following the advice to "dream big." There, a seed was planted, and as a young kid, I went off chasing the idea that I could do what I dreamed to do in a world I knew nothing about. While there was some truth to that idea, the path was not a straight line. Instead, it was a twisted road up a mountainside toward a very high summit. From Bible College to Hollywood, from sailing the seas to producing National TV, and from Italy to New York, I'd followed my path, and now it seemed like it had been all divine.

Wrapping up my last day in India, I was preparing to go when I learned that the Dalai Lama was speaking in an open field a few miles away. Jumping in a car, I drove down to find thousands of monks in amber-red robes and bald heads sitting cross-legged, listening to him speak. As I was ushered to the small tourist section on the side, I was in awe of what I was seeing. Only in the realm of a mythical film would a scene feel so majestic and so surreal as this. But this was not a film; this was reality. Sitting and listening through the earbuds to the translation, I was uplifted by the message. It was a special end to an amazing voyage.

As I boarded the plane to return home, the future began to infringe on the present. I recalled that I was again unemployed

CHAPTER 12

and living in one of the most expensive cities in America. Sinking into my seat, I again faced the realities of the unknown, and fear again began to creep in. While a part of me wanted to charge boldly into the future, applying the faith that worked so hard to practice, another part of me doubted everything I believed and wanted to run away and hide. It was the painful irony of the reality of life and a very applicable example for needing the Buddhist motto: "After enlightenment, chop wood and carry water."

Stefania & Barry at there second wedding on the beach outside of Rome.

The screening of *The High: Making the Toughest Race on Earth* at the Main Art Theater in Royal Oak, MI (2015).

"It's not where you're from; it's who you become."

—Rita Levi-Montalcini

CHAPTER 13

HOME AGAIN FOR THE FIRST TIME

NEW YORK TO DETROIT (2012)

On my return to New York after filming in India, the first thing on my mind was to find a new job, but after nine months of living near the city, I questioned if this was the right place. Commuting three hours a day from our home in Long Beach to Manhattan was a lot, the cost of living was high, and the pace of life was faster than I wanted. While I loved the energy of the place, I had a strong desire to start a family, and with that desire, I thought of my childhood home in Michigan.

No sooner did I think of returning to the motherland than a position became available to work for the Detroit Pistons. I couldn't believe it. I had grown up with the Detroit "Bad Boys" as a kid. I watched Isaiah Thomas, Bill Laimbeer, Dennis Rodman, and more win back-to-back championships in 1989 and 1990, all at The Palace of Auburn Hills. In addition, basketball as a sport had been my first love going all the way back to childhood. It harkened back to the dirt driveway in Medina. For me, this was a dream job and come hell or high water I was determined to go get it. As quick as a whip, I scrambled together a résumé and sent it in, then started looking for people who could recommend me for the role. To my surprise, two weeks later, I got an email from the HR department wanting an interview. I was instantly on a plane heading back to the Motor City. After an interview in Detroit and winning CEO Dennis Mannion over, I was offered the position of working on the courts of the NBA. In October 2012, just two months after returning from capturing *The High Race* in India, I moved my Italian wife from Rome, Italy, to the suburbs of Detroit—back to the land where it had all started.

LESSONS FROM MY JOURNEY

One of my favorite lines in life is a quote from T. S. Elliot's poem, "Little Gidding," in which he wrote, "We shall not cease from exploration and the end of all our exploring will be to arrive where we started and know the place for the first time." No words could better sum up what I felt upon reentry into Michigan and our new home in Detroit.

For the past twenty years, I'd traveled into the unknown, hoping to filter through the chaos for answers to who I was and why I was here. During that time, I'd found that everything I needed to know had always been inside me. In my search to fulfill my dreams, achieve goals, and answer the questions of life, I discovered that one does not need to go further than their own doorstep to uncover everything there is to understand about life. This philosophy is echoed in Hermann Hesse's novel, *Siddhartha*, a story about a man who learns everything there is to know about life from sitting under a huge bodhi tree next to the river. I realized that everything I'd felt I was missing as a child always existed within. I had been miraculously created and was complete, even in my shortcomings, failures, and flaws from the day I was born. I only needed to see it for myself. For the first time since my return, I could see this, and it felt good.

CHAPTER 13

My choice to go in search of myself, however, did come with a great reward. Going out beyond the walls of my childhood drove me on a path of self-discovery, and that path taught me some of the most important lessons in life. From miles on the road, in the air, on planes, and at sea, I uncovered a practical understanding of how to turn dreams into reality. I grew a deeper understanding of the limiting power of fear and discovering in the process the value of clarity. I developed a practical application of faith. And I learned the necessity of having balance in life, learning to walk the path between chaos and order, yin and yang, and the known and the unknown.

THE FINISH LINE AND THE CAVE

Now living in Detroit, I was technically back home. After being settled, I traveled with my wife back to our country home in the small town of Medina, where I grew up. Standing there, looking at that beautiful white house on a hill, it was clear time had passed and things had changed. My parents had long moved away, the woods had grown bigger, and the trees older. As I began to reimagine those days so many years ago, I could see myself as a little boy who loved to dream big—possibly too big for this little town. Amazed to be here, I realized that I'd come full circle, and that little boy, with his fear and insecurity, was only here now in spirit; his presence had left this place long ago.

Back in our home outside of Detroit, I was editing on the film I'd shot in India. In front of me, in the colorful lines of my software, were millions of clips piecing together a story of overcoming the impossible. In the cold of a Michigan winter, working in my basement office next to my space heater, I enjoyed the feeling of escape to the high altitude of the Himalayas. While working on the pieces of the film, somewhat humorous delusions of grandeur filled my thoughts, and I imagined being lifted up on the crowd's shoulders, cheered for my great work, and celebrated by my family and friends for making such an amazing film. It was the silliness of thought that drove me to keep editing for hours on end.

In the spring of 2013, with tulips pushing up through the melting snow, Stefania announced she was pregnant, and everything else in life became secondary. I was going to be a father, and something far greater than any film was soon to be born. On November 30th, 2013, in Rochester, Michigan, my son David Christopher Walton arrived. It was the greatest day of my life.

Completing my work after nearly five years with the help of some great friends, I rented out and screened *The High: Making the Toughest Race on Earth* at The Main Street Theater in Royal Oak, Michigan. On a Thursday night, the lights of the projector flickered over the heads of 300 guests. There, I was both humbled and thrilled as I watched my film play on the silver screen. It was another great accomplishment and one of the greatest nights of my life.

In the years following the completion of *The High*, I started work on the footage from the sailing documentary that I'd filmed some fifteen years prior. In August 2019, I was accepted to screen the film at the Santa Cruz Film Festival in Santa Cruz, California. In a twist of fate, I again had come full circle and was joined by a reunion of my pseudo family of friends from LA to watch my film *Reaching Reality: A Sailing Adventure of Making a Dream into Reality*.

In years to come, *The High*, its sequel *Ultra High*, and *Reaching Reality* have been available on Amazon Prime and enjoyed by hundreds of thousands of viewers to watch around the world.

THE CAVE YOU FEAR

In the days leading up to the start of writing this book, I had a revelation. In the dust of boxes and family archives in my parents' basement, I picked up a thinly bound book I'd written in the fifth grade entitled *The First Boy in Space* by Barry Walton. The book was written in my best penmanship and illustrated

CHAPTER 13

by me. Sitting there, flipping through the pages, I noticed on its cover was a golden sticker representing that I'd won first prize for best book of the year. For the first time in my life, I realized that there in front of me was one of the greatest achievements in elementary school. In one of the most emotionally painful periods of my life, my early education, where so much had been misunderstood, I had found a silver lining. Amid my failures and disappointments, I had written a book that had won top honors.

Holding this personal archive in my hand, I realized it was an opportunity to change the narrative about my past. It was an opportunity to give life to a new vision for my future. Then and there, I decided that instead of seeing all my school years as a failure, I would see it as a process toward success, and this book was a shining star in that work. Those years have made me what I am and have always been—a storyteller, a writer, a filmmaker, an adventurer, and a traveler into the unknown.

In reshaping my story, I could see that history was no longer a burden that I needed to bear. Instead, it was a blessing. Alone on that day of self-realization, I put down that wonderful reminder of my past, returned to my keyboard, and started to type the book you are reading today. The finish was now the start, the problem was now the solution, and the cave that I had most feared to enter had always held the treasure that I sought.

My first book from 4th grade, "The First Boy In Space"

LEARN MORE ABOUT THE AUTHOR

In the process of turning your dreams into reality, go to EndlessMedia1.com to find films, social media links, and upcoming speaking events with Barry Walton.

www.ingramcontent.com/pod-product-compliance
Lightning Source LLC
LaVergne TN
LVHW020400250125
802081LV00001B/125